*Race, Class, and Power in the
Alabama Coalfields, 1908–21*

THE WORKING CLASS IN AMERICAN HISTORY

*A list of books in the series appears at
the end of this book.*

Race, Class, and Power in the Alabama Coalfields, 1908–21

BRIAN KELLY

University of Illinois Press

URBANA, CHICAGO, AND SPRINGFIELD

⊗ This book is printed on acid-free paper.

Library of Congress Cataloging-in-Publication Data
Kelly, Brian, 1958–
Race, class, and power in the Alabama coalfields, 1908–21 / Brian Kelly.
p.   cm. — (The working class in American history)
Includes bibliographical references and index.
ISBN 0-252-02622-5 (alk. paper)
ISBN 0-252-06933-1 (pbk. : alk. paper)
1. Coal miners—Labor unions—Alabama—History. 2. Afro-American
coal miners—Alabama—History. 3. Afro-American labor union mem-
bers—Alabama—History. I. Title. II. Series.
HD6515.M615K45   2001
331.88'16262'0976109041—dc21   00-009895

Contents

Acknowledgments

Had I been a more superstitious person, or even mildly inclined to put much faith in omens, I would have abandoned this study at its inception. On the basis of a few scraps of intriguing evidence and with some much-needed financial help from the graduate history program at Brandeis University, I set off from Boston during a January break some years back to visit archives in West Virginia and Alabama. Less than an hour into the trip it began to snow, and by the time I reached New York I was caught in the worst blizzard of the decade. For most of the second leg of the trip I was obliged to meander along, uncharacteristically, at somewhere under half the speed limit, and as I observed my fellow drivers careening off of guardrails and vanishing over the highway embankment I wondered if it was all worth the effort, not to mention the risk. Things did not improve much upon my arrival at Morgantown. Weather closed the library at West Virginia University for three consecutive days, and as my modest stipend began to dwindle under the strain I was reduced to buying a snow shovel to keep myself solvent. The snow fell so hard and for so long that by the third day I was doing enough shoveling business in Morgantown's commercial district to extend my stay in town.

The storm eventually ended, the university returned to normal, and I found the answer to my earlier doubts during my first few hours on campus. Two things in particular relieved whatever misgivings I'd entertained about the project. The first I found in the archives. Buried in the UMW organizer Van Bittner's papers at Morgantown were two or three hundred letters sent to the union's headquarters by miners and their wives during the course of the 1920 strike detailing the determination and solidarity that sustained them and the inhumanity to which they had been subjected. Theirs was a story that deserved to be told, and if I have conveyed in the pages that follow something of what moved me in those compelling documents then I will be satisfied. The other spur for pushing ahead with this study came later that day when I sat down with Ronald Lewis, then the chair of the history department at WVU and the author of a seminal study on African-American coal miners. Generous with his time, helpful with source suggestions, and enthusiastic in his support for the project, Lewis played an important role in encouraging a somewhat bewildered graduate student in the early stages of a still vague and undefined research project to persevere.

His support came at a critical stage in this project, but my debt to Lewis was neither the first nor the last that I would incur during the time it took to see it through. Over the course of my research and writing, I've been fortunate enough to have had the encouragement of a number of extremely gifted historians and the support, financial and otherwise, of several key institutions. I owe my deepest debt of gratitude to Jacqueline Jones, who supervised the dissertation upon which this study is based and who has followed it through to publication, and to the history department at Brandeis University, which for several years generously supported my annual pilgrimage south to the archives. Graduate study would have been impossible for me if I had not been the recipient of a Crown Fellowship while at Brandeis, and I am grateful to everyone involved in affording me the opportunity to return to study after an extended sojourn in the construction industry.

I have also benefited from ongoing discussions with scholars working in the fields of labor, southern, and African-American history, many of whom have contributed in one way or another to this work. They include Alex Lichtenstein, Peter Alexander, Eric Arnesen, Ed Brown, Peter Cole, Colin Davis, Gary Fink, Wayne Flynt, Jim Green, Michael Honey, Daniel Letwin, David Lewis, Robert McMath, Kevin Murphy, Kimberley Phillips, Leslie Rowland, and Judith Stein. Rick Halpern and David Montgomery read the manuscript for the University of Illinois Press and offered thoughtful, constructive suggestions for improving it. Jeff Jakeman and his staff at the *Alabama Review* pushed me to clarify my work in its early stages, a process that was helped along by the opportunity to present work in progress before the Southern Labor Studies Conference, the Southern Historical Association, the Auburn University Symposium on New Perspectives in Southern History, the London Seminar in Comparative Labour History, and the North American Labor History Conference.

As every working historian knows, scholarship is in many ways a collective enterprise. At every stage of this project, library staff were invariably helpful and generous with their time and enthusiasm. I especially want to acknowledge the help of reference librarians at Brandeis University and of archivists at West Virginia University's Regional History Collection, the Department of Archives and Manuscripts at Birmingham Public Library, and the Alabama Department of Archives and History at Montgomery. Three institutions share credit for keeping me gainfully employed while finishing the study, and I am grateful to colleagues at Suffolk University, Florida International University, and Queen's University in Belfast for their support. Thanks especially to Bill Walker and Elena Maubrey at FIU. Tula Petridou translated several key documents from Greek; Lisa Long of the Farber Archives at Brandeis helped track down some of the photos that appear in the pages that follow; Lenora Canon gave me access to her father's collection of interviews with rank-and-file miners; Craig Remington of the University of Alabama Cartography Lab assembled the map; and Constance Jones-Price talked with me for hours about the perils endured by her father, UMW organizer Walter Jones. Richard L. Wentworth at the University of Illinois Press has been a source of unstinting encouragement, and Theresa L. Sears has navigated the manuscript through to publication. Matt Mitchell and Jim O'Brien were meticulous, respectively, in copyediting the manuscript and compiling the index. My good friend Michael Pearce proved a credit to his hometown during my frequent visits to Birmingham.

Like the work of many labor historians, this study was motivated by something more than a purely academic curiosity about a series of events in early twentieth-century Alabama. It is informed by a long-term engagement with the labor and socialist movements in which I first consciously aspired to understand the world. I'm grateful to all of the many friends and comrades who've helped me in that pursuit over the years. Through thick and thin, all eleven of my brothers and sisters have been there for me throughout this project, and I wish to express my profound gratitude to each of them. Finally, I want to dedicate this book to my daughter, Cíara, as a modest return on the joy she has brought me these past sixteen years, and to my parents, Ed and Mary Kelly, who have seen us through.

Race, Class, and Power in the
Alabama Coalfields, 1908–21

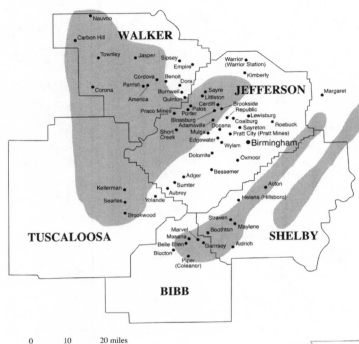

Major Coal Mining Towns
of the
Birmingham Mineral District:
1900-1920

Introduction

Nearly seventy years ago, the renowned scholar and activist W. E. B. Du Bois laid bare the central paradox of southern labor history. The episodic attempts on the part of poor black and white southerners to collaborate across the color line, marked as they were by hesitation and distrust and driven more by desperation than conscious egalitarianism, had been continually frustrated. But the difficulties of breathing life and continuity into working-class interracialism, Du Bois insisted, were neither predetermined nor inevitable. Though the twin legacies of slavery and race prejudice continued to weigh heavily upon the region well into the twentieth century, the obstruction of class solidarity in the American South rested not so much upon cultural immutability as upon a dynamic and deliberate intervention.[1] In the resurgence of popular racism activated during the overthrow of Reconstruction, Du Bois discerned a more or less concerted plan of action on the part of white elites. The deeply rooted racial phobia that saddled every stratum of the white South "had been supplemented by a carefully planned and slowly evolved method" directed from above. The resulting antagonism, encouraged by the South's most powerful agricultural and industrial interests, "drove such a wedge between . . . black and white workers," he wrote, "that there are probably not [anywhere] two groups of workers with practically identical interests who hate and fear each other so deeply and persistently that neither sees anything of common interest."[2]

Du Bois's keen if sobering observations about the debilitating effects of racism on working-class organization have become something of a touchstone in contemporary discourse about the relative importance of, and the tangled relationship between, race and class in American history. And so they should be. But scholars have been conspicuously selective in choosing which elements of Du Bois's perspective to take on board. One side of his formulation, at least, seems uncontroversial. An outpouring of imaginatively conceived, solidly researched scholarship on black workers and their relation to the labor movement over the past twenty years has reaffirmed the notion that the fundamental divide among American workers has been and continues to be the color line.[3] Nor are new excavations into the past likely to unearth evidence to the contrary. Despite sharply divergent approaches to the study of black workers and their relation to organized labor, few scholars would dispute either the historical persistence or the destructive power of racial antagonism in the American past.

The other essential element in Du Bois's conception of the problem, however, which emphasizes the deliberation with which racial divisions have been maintained, has fallen out of fashion in recent years. Contemporary scholars seem more inclined to dismiss as conspiratorial or "reductionist" Du Bois's insistence that racial antagonism at the bottom of southern society had been "carefully planned," that "Southern capital accepted race hate and black disfranchisement as a permanent program of exploitation." Indeed, an argument that would have formerly passed as common coin among those who either chronicled or attempted to make southern labor history—that white elites and not white workers were the main beneficiaries of racial division—is explicitly rejected in much recent writing on race and labor. Practically eliminated from contemporary scholarship is the working understanding of racism embraced by an earlier generation of left-wing labor activists as a "policy of divide-and-rule" by which the "well-to-do planters and the southern ruling class generally . . . derived a benefit from the struggle that black and white workers were carrying on between themselves." In its place, academics today hold up an explanation in which employers' motivations figure hardly at all, in which the notion that joint struggle around shared material interests might overcome racial antipathy is dismissed as quaint, and in which a timeless and irreducible racial identity—"whiteness"—inevitably trumps the feeble potential for interracial solidarity.[4]

In part, scholarly aversion to a materialist explanation of racism[5] is one of the collateral, though unexpected, developments that attended the rise of the "new social history." Determined to move beyond the top-down, institutional studies that had dominated their field since its inception,[6] labor historians swept up in the social and intellectual ferment of the 1960s committed themselves to rewriting working-class history "from the bottom up," convinced that such an approach would yield vast new insights. And for the most part they have delivered the goods. Transforming a field that rarely delved beyond descriptive accounts of the confrontation between the "lieutenants of labor" and their counterparts in the boardrooms of organized capital, labor scholars have extended their research to produce richly textured studies that treat seriously organized and unorganized workers; slaves, wage-earners and their families; black, white, and immigrant workers; men, women, and children. They have moved out from the point of production to examine both the public and overtly political and the private, domestic spheres of workers' lives, churning out studies that show little respect for the boundaries accepted by their forerunners.

From the beginning, however, even scholars who were exhilarated by the expansion of their field voiced reservations that while the new social history cleared away barriers to a more comprehensive exploration of working-class life, the view "from below" might merely substitute a new form of historical myopia for that of the "old" labor history it aimed to displace. In particular, their theoretical agnosticism and enthusiasm for history "with the politics left out" meant that the "veritable cornucopia of information" being harvested by the new labor historians was of "rather uncertain usability." Writing in 1979, David Brody welcomed the new turn but cautioned that the "narrow focus of our research, our devotion to intensive, local study of workers," and the "acute sense of the complexity and variety of working-class experience" brought to bear in new studies militated against the elaboration of a new synthesis. Less than a decade after the field's celebrated departure from the "safe haven" of the institutional framework, Brody noted the stubborn persistence of familiar problems related to power and organization, formal politics, and conflict at the point of production. Perhaps what the field had gained in texture and detail, some warned, had been won by sacrificing a sense of the larger context of workers' lived experience.[7]

Brody's ambivalence reflected growing unease at the new social history's tendency to obscure rather than clarify the relationship between working-class activism and the "facts of power" in modern industrial society.[8] Elizabeth Fox-Genovese and Eugene D. Genovese issued a trenchant, if acrid, critique of the new turn, characterizing its "fascination with custom" and "tendency to look to private existence to redress the political impotence of the public" as an "attempt to put everything and anything . . . in place of . . . the fundamental problem of power and order." After all, they asserted, "history . . . is primarily the story of who rides whom and how." From Britain, John Saville articulated similar concerns, objecting that "the study of working-class institutions and movements" was too often being "undertaken in isolation from the rest of society." The result, he warned, was a notable retreat from historical generalization and a "failure to situate the working class within the totality of social relations." Reasserting the legitimacy of "total history" against the postmodern "cult of the fragmentary," Alex Callinicos has more recently insisted that while there remains "much to . . . defend, and indeed to celebrate in the 'history from below' tradition," its explanatory potential is blunted by an aversion to "study[ing] the organization and strategies of the dominant class as well as the life and struggles of the oppressed," leaving the new scholarship vulnerable to criticism from those who deny the very possibility of reconstructing the past.[9]

Though its negative results were unforeseen, popular determination to restore agency to those marginalized in earlier accounts of the past reinforced the possibility that social historians would bury the main lines of American labor history under a mountain of seemingly random and unsifted detail. Like the reaction against the institutional approach of the old labor history, the impulse was in many ways a healthy response to the near-universal ascendancy of traditional, elitist "history from above." Guided by a method that tended to dodge the centrality of power, however, the endeavor never fully confronted the most salient question it evoked: how does one meaningfully measure the agency of social groups that were, relatively speaking, powerless? From her vantage point at the cutting edge of the new western history, Patricia Nelson Limerick identifies a conceptual dilemma that every field affected by the revolution in historical writing over the past thirty years will have to confront: "We have become so ardent in addressing agency, in ferreting out all the ways in which . . . oppressed people still managed and directed their own destiny," she writes, "that we have nearly silenced ourselves when it comes to . . . the ways in which they were taken advantage of and overpowered."[10] In an undertaking that deciphered agency in the most obscure corners of people's daily existence, the relative disparity in power between working people and elites, which an earlier generation of labor historians had taken for granted, forfeited its significance as an essential starting point for unraveling working-class history.

For a variety of reasons, then, the outlook of the new labor historians predisposed them against attaching too much to Du Bois's contention that the antagonisms at the bottom of southern society were "carefully planned," that black and white workers had been "kept . . . far apart" by southern elites with a material stake in racial disunity. Among another emerging school of labor historians and political scientists, however, the rejection of Du Bois's perspective became a matter of principle. The soft-pedaling of capital's role in fostering and perpetuating racial divisions developed as a necessary corollary to the celebrated Gutman-Hill debate. Herbert Gutman had suggested, rather tentatively, that alongside "evidence of racial mistrust and antagonism between whites and blacks" there existed strong evidence that—in some industries and in some locales, at least—relations between black and white workers had been characterized by genuine interracial cooperation. While lamenting the "absence of detailed knowledge of the 'local world' inhabited by white and Negro workers," Gutman noted the "successful early confrontation between the United Mine Workers [UMW], its predominantly white leaders and members, and Negro workers" and maintained that the UMW's championing of interracialism made it "the most thoroughly integrated voluntary association in the United States of 1900."[11]

Gutman's exploration of the early career of the black UMW organizer Richard L. Davis was harshly criticized some twenty years after its publication (and three years after the author's death) by the former NAACP legal adviser Herbert Hill as an exercise in "myth-making." Rejecting the "myth of the UMW's benevolence towards the black worker," Hill posited an ideological explanation for the alleged defects in Gutman's scholarship. The "tendency to deny race as a crucial factor, to permit questions of class to subsume racial issues," which he detected in the Davis article, was ascribed by Hill to the lingering influence of "a romanticized 'popular front' leftism that became a major characteristic of the Gutman school."[12]

While the general contours of the ensuing debate have been recounted too often elsewhere to justify a detailed rendering here, one aspect of Hill's argument seems relevant to the problem of Du Bois's legacy. Gutman's account of Davis's organizing career did not deny the existence of sharp racial divisions in the UMW. He projected that "any authoritative history of the UMW" would necessarily "include grimly detailed pages about racial and ethnic quarrels and even death and violence" alongside its laudable, if complex, record of interracialism. But while Gutman considered it worth pointing out that the "coal operators often exploited tensions between Negro and white miners" (quoting Davis's complaint that they "'keep up a distinction between men for the purpose of breeding strife and dissension in our ranks'"), the employers disappeared completely from Hill's account. Focusing rather narrowly upon the prevalence of racism within the UMW itself, Hill cited an earlier study which noted that the union "[had] not obtained equal treatment for Negroes in the allocation of job opportunities." But in doing so he neglected the obvious fact that coal operators themselves had exercised far more control over the racial division of labor than the UMW ever enjoyed. In Alabama, where the UMW fought against the convict lease and the notorious subcontracting systems (both of which weighed heaviest upon black miners), the elusiveness of equal treatment for blacks had less to do with white miners' racism than the union's inability to match the operators' awesome power.[13]

In recent years this peculiar blind spot for organized capital's role as an active participant in the construction of race has been welcomed, by some, as a positive advance over earlier, marxist-influenced models. Hill has recently argued against the ostensibly dogmatic notion that employers engaged in a "divide-and-conquer strategy" in dealing with racially mixed groups of workers and appears at one time to have looked to enlightened southern employers to lead the fight against segregation.[14] In *Iron and Steel: Class, Race, and Community in Birmingham, Alabama, 1875–1920*, Henry McKiven rejects "the argument that employers [used] white supremacy to blind white workers to their long-term interest in uniting with [blacks]," suggesting instead that white Birmingham steelworkers "considered themselves the equals" of the men who employed them and, through their unions, "played a primary role in . . . constructing an ideology of white supremacy to secure and . . . justify their power and status." But the extent of that power was apparently lost upon contemporary observers (including employers), who commented almost universally upon the weakness of union organization in district mills.[15] Similarly, the political scientist Allan Draper's study of southern white labor's hostility to civil rights, much of it based on Alabama sources, depicts a hard-pressed union bureaucracy mounting a valiant defense of black equality against a reactionary white membership. Draper regards the prevalence of racial hostility among whites as a natural expression of their "class privileges" but conspicuously ignores the panicked, often violent opposition of Birmingham employers to every previous manifestation of working-class interracialism. To the extent that capital enters his account at all, it does so as the host of racial reform rather than a bulwark of division.[16]

Under the guise of exorcising a marxist teleology (real and imagined), the Hill school, its stature enhanced by new cultural studies that stress the primacy of racial identity—"whiteness"—in the very formation of white workers' class consciousness, has constructed a framework that leaves one wondering whether white supremacy served any function *other* than defending the material and psychological interests of working-class whites.[17] In place of the more optimistic perspective that animated black and white union activists during American labor's last major crusade in the 1930s, "the pendulum has swung" quite dramatically in the direction of "a Right essentialism that considers the white worker to be inherently and inescapably racist."[18] At times these accounts convey the impression not only that southern capital did not benefit from Jim Crow but that employers chafed under the system, assenting to it only to mollify their white minions.

My own reading of the tumultuous struggles that convulsed the Alabama coalfields in the early years of the twentieth century points in a very different direction. Fundamentally I argue that the persistence of race hatred cannot be meaningfully understood outside the framework of class relations imposed upon the industrializing South by regional elites. Du Bois was not wrong to insist upon the connection between Jim Crow and industrial development in the New South, nor was his a trivial observation. Birmingham district coal operators might have been willing, in rare cases, to relinquish the more ornamental elements of the South's racial order. Small operators who could not afford to build separate housing clusters for black and white employees may have harbored frustrations that the costs of southern racial protocol were excessive. They and their more powerful rivals may have regretted the unpredictability introduced into their operations by intermittent, socially sanctioned racial violence. But they were absolutely clear, like their counterparts in steel, that white supremacy was essential to maintaining their hold over what one operator termed the "'greatest, best and cheapest labor market in the United States.'"[19]

Leading operators, many of whom considered themselves agents of progress and bona fide "friends of the Negro," were nevertheless dependent upon the continued subjection of those they employed. Race prejudice did not merely linger like some genetically induced hangover in the mind of the twentieth-century white South; instead it acquired a new lease on life in the region's scramble for industrial prosperity. Racial subordination and industrial exploitation were not discrete systems of social control operating independently in a haphazard fashion. Together they formed an organic and indivisible whole, constituting the cornerstone of the region's plans for development and, ironically, the key element in its formula for "progress."

This study advances an argument for "bringing the employers back in" to the long-standing debate over race and the labor movement. It proceeds from the assumption that any framework that omits from consideration the central actors shaping the terrain upon which black and white workers interacted is intrinsically flawed. Alabama operators were conscious of—even obsessed with—the strategic import of race throughout the period under consideration. At the nadir of UMW influence in the years following the 1908 strike, they patched together a system of labor relations that grafted the most advanced model of labor management in the world onto more traditional methods rooted in the region's slave past. In the coal camps and mine villages scattered throughout the Birmingham district an ambitious reform program inspired by northern-based welfare capitalism coexisted with convict labor, the company "shack rouster," and the whip. The centerpiece of the operators' project was a system of racial paternalism that aimed to take advantage of existing racial divisions and the relative vulnerability of black workers under Jim Crow to erect a permanent barrier against unionization.[20] This strategy imposed a delicate balancing act upon area operators, one that demanded that they uphold an enthusiastic commitment to white supremacy even while flaunting their credentials as the natural allies of the black miner and that they maintain the allegiance of black employees even as they went about fleecing them ruthlessly.

Such tortured public posturing could be maintained only in the absence of an effective mineworkers' organization. The operators' system began to crack, however, and finally to collapse, under powerful pressures brought on by World War I. From 1917 onward they faced a renewed interracial challenge, culminating in the district-wide coal strike of 1920–21. The employers' panicked reaction to the revival of interracial unionism, culminating in the resurrection of Klan-style vigilantism and military suppression of the UMW, reveals much about the essence of white supremacy in the industrial South that a narrow focus on workplace racial "identity" would otherwise miss.

The alternative to the Hill school's positing of a fixed, overriding attachment to racial supremacy on the part of white workers is not a sanitized rendering of events in which hostility is swept under the rug or rationalized as the inevitable manifestation of employer manipulation. White miners do not emerge from this study as egalitarian knights, but they do emerge as a more complicated and varied mass. Though deeply inscribed by the pathological culture in which they developed, their ideas on race were subject to change in the protracted social crises that industrial confrontation produced in the Birmingham district. Their attachment to white supremacy, though never completely transcended, was nevertheless fundamentally shaken, both by their treatment at the hands of the "better elements" of "their" race and the necessity of linking arms with black miners. Hill is almost certainly correct when he argues that white workers typically opened their unions to blacks out of necessity rather than racial altruism. But he evades the more fundamental question, raised by Stephen Brier, of what transpires when "white miners, whose attitudes if not their behavior towards blacks are on the whole racist, begin to shed practices as they begin to participate in inter-racial organizations and actually see their fellow black workers as active union supporters."[21]

This study shares Brier's insistence on the dynamic quality of these interactions, on examining the experience of interracialism, and assumes the more fundamental responsibility of gauging the outlook embraced by Alabama miners against the ideas that dominated their society. I have taken seriously John Saville's understanding of labor history as "the story of the working class within a society whose social parameters have, in the main, been determined not by themselves, but by other social factors and classes." Alabama miners "made their own history," to borrow a phrase, but they did not "make it just as they please[d] . . . under circumstances chosen by themselves."[22] Although at the height of UMW power miners succeeded in registering their opposition to the vast inequalities afflicting southern society, always and everywhere their own maneuvering was limited by the overwhelming social, economic, political, cultural, and, yes, military hegemony of their employers.

Numerous studies have identified the wartime and immediate postwar years as a watershed in both African-American and working-class history. The crisis of southern agriculture that preceded the outbreak of war spurred the first mass migration of blacks out of the South, a process that intensified with military and industrial mobilization and laid the basis for the first major opening for black self-assertion since Reconstruction. Although it has been too often overlooked, the defining feature of these developments was the entry of the black working class onto the stage of African-American history. Judith Stein's seminal point, that "organized [black] workers altered the racial agenda" in the immediate postwar period, that wartime transformations instilled among black workers the confidence to challenge the authority of conservative, prewar race leadership, is confirmed in the pages that follow. Even where he or she did not directly impinge on struggles at the point of production, the "new Negro" that struck fear in the hearts of white racists and black conservatives alike was by and large a proletarian.[23]

The same charged atmosphere that spawned increased militancy among African Americans produced heightened expectations among American workers in general, and the end of hostilities in Europe opened up a period of remarkable confrontation between labor and capital at home. Working-class militancy, barely contained during the war itself, burst forth in a massive strike wave and met with determined opposition from government officials and employers across the United States, who resuscitated the apparatus that had served them so well in a similar crisis more than forty years earlier: heavy-handed military repression, anti-immigrant xenophobia, and bellicose antiradicalism.[24]

Although its importance during the period has been underrated, Birmingham was one of several epicenters where these two dramatic, high-stakes confrontations converged. Nationally the outcome of these clashes varied depending on a number of contingent factors; more often than not the result was bitter, bloody confrontation between white trade unionists eager to reap the fruits of their patience during the war and black workers anxious to win a foothold in industrial employment from which they had been long excluded. In some places, most notably the Chicago packinghouses, the sto-

ry was one of hopeful beginnings—groping attempts at interracial cooperation—followed by a sharp descent into racial brutality.[25] But in the coalfields
of northern Alabama these two powerful currents came together, and held
together, in the face of determined attempts by the employers to pull them
apart. The story of the bitter 1920–21 Alabama coal strike, narrated in detail
in the final chapter of this study, is an important episode in a broader national plot of "decisive confrontation between the [American] working class
and the state" in the immediate postwar period.[26]

The search for an explanation of this remarkable resiliency merits a detailed treatment of wartime militancy in the coalfields. No one who has had
the opportunity to absorb Daniel Letwin's powerful rendering of the forma-
tive period in Alabama coal unionism will envy my own position in setting
forth some conclusions.[27] A model of clarity, precision, and even-handed-
ness, *The Challenge of Interracial Unionism: Alabama Coal Miners, 1878–1921*
showcases the very best elements in an emerging reappraisal of southern labor history, one in which "little questioned opposing formulas—class vs. race,
solidarity vs. fragmentation, white unions vs. black workers—are coming
under close critical scrutiny."[28] Letwin carefully delineates the remarkable
transformative potential of coalfield interracialism and its acquiescence in
the face of deeply embedded racial custom, while keeping an eye on the shifting motives of district operators.[29]

Insofar as this study focuses upon racial dynamics within the UMW itself,
it complements rather than challenges Letwin's interpretation. But the non-
union period that this study focuses upon illuminates several important
problems underemphasized in Letwin's expansive study, and on two vital
issues the sources lead me to significantly different conclusions. First, I as-
cribe to the operators a more fundamental and consistent role in maintaining racial divisions than Letwin's study seems to suggest. In the period before 1908, he finds area operators too preoccupied with patching together a
stable workforce to pursue the strategic advantages that might be gained from
playing the race card. Prior to 1893, Letwin writes, there is "no evidence that
the operators' recruitment of both blacks and whites represented any sort of
divide-and-rule strategy." Their ambivalence gave way, however, during the
1894 strike when, led by the district coal pioneer and staunch antiunionist
Charles DeBardeleben, operators deliberately "adopted bold new strategies
to divide the miners along racial lines." Tensions subsided again, according
to Letwin, in the period of accommodation that straddled the turn of the
century. During a brief interlude area operators seemed reconciled to a union
presence and refrained from raising "the inflammatory race issue, the point
on which the union was most vulnerable." But this era of relative "good feel-

ings" was short-lived. Anxious to reassert its authority in the mines, TCI management led the operators into a new series of confrontations with the UMW beginning in 1903—clashes in which race figured prominently.[30]

The strength of Letwin's account is his insistence upon the mutable quality of the operators' public posture regarding race: in general they resorted to open race-baiting only when faced with a direct challenge to their authority in the mines. But by itself such an interpretation understates the continuity in the operators' obsession with race. The assembling of a racially mixed mine workforce in the years immediately following Redemption was not merely a matter of securing sufficient labor but of putting together a workforce that was sufficiently tractable. Operators brought to this process a range of racial assumptions that imposed upon the mine workforce two alternatives: either black labor would provide the operators with a powerful lever with which to hold down the wages and conditions of all miners in the field, or black miners would be organized on more or less equal terms with whites to deny the operators that possibility. Regardless of whether operators engaged at any given time in open racial demagoguery, the potential for racial conflict was latent in the strategy they pursued from the beginning. This is the subterranean thread that runs from the late 1870s through the period covered by this study. In some sense the distinct form that welfare capitalism assumed in the district coalfields after 1908 marks the high point in the operators' pursuit of this strategy, and the renewed interracial challenge brought on by the war represents the zenith of miners' resistance.

As Letwin's study confirms, black miners were neither the quiescent objects of employer paternalism nor the mere pawns of white trade unionists. Instead they played an active role in defending their union and pushing it to live up to the spirit of its commitment to equality within the ranks. Ironically, the sternest critics of the new southern labor history—Hill, McKiven, and Draper among them—have been the least successful in integrating black workers as subjects of their own history. Such a glaring defect cannot be attributed to a lack of sources. Black workers left few direct records of their own sentiments, but their perspective can be reconstructed by trawling deeply through a range of sources.

After the turn of the century, the stabilization of Birmingham's black community gave rise to a more or less permanent black press, leaving a paper trail that documents an increasingly intimate relationship between white employers and black elites. The sources become richer again during the period of the First World War, when various federal agencies directed their attention to the "problems" of black emigration, the disruptive potential of militant black nationalism, and the specter of revived industrial militancy in the nation's factories, mines, and mills. Finally, the strategic importance of Birmingham as a nerve center of black southern cultural and political life made it the focus of organizing efforts by rival race tendencies attempting to challenge the authority of Booker T. Washington's Tuskegee machine. Collectively these sources suggest that the long-standing debate over protest versus accommodation, which has focused almost exclusively upon differences among black elites, must be recast to take account of the distinct if incipient perspective of black workers who, after all, comprised the overwhelming majority of those languishing under Jim Crow.

The most significant casualty of the reconstruction of black Birmingham presented here is the "'sentimental notion of black solidarity'" that has long dominated studies of the Jim Crow South.[31] The conceptual subordination of class to identity among academics in recent years has reinforced a long-standing taboo against sustained investigation of intraracial conflict, but the tensions that this study reveals within Birmingham's black community affirm the necessity of a break from outdated assumptions. The coexistence of an intermittently viable interracial UMW alongside a conservative black middle class steeped in the accommodationist tradition made for an explosive confrontation among African Americans themselves. The outlines of the conflict between black miners who, when afforded the possibility, linked their fortunes to those of white UMW members and "race leaders" who argued stridently for an alliance with the "better class of white men" are clearly discernible in the documentary record. Here again I draw conclusions different from Letwin's: far from representing an example of the compatibility between racial accommodation and black working-class militancy,[32] this study suggests that the Alabama coalfields might well have provided the Tuskegee machine with its most significant challenge anywhere in the South.

Finally, although this study makes no claim to the universality of the Birmingham experience, I argue that it does bear implications beyond Birmingham. If the Alabama mining district constituted an anomalous feature on the industrial landscape of the New South, it was not because the conditions endured by black and white miners were exceptional: laborers in the region's timber and turpentine camps, iron ore mines, docks, levees, and even cotton fields lived and worked under regimes that would have felt familiar to most miners. Birmingham's exceptionalism lay in the fact that its coal miners—unlike most of their counterparts elsewhere in southern industry—succeeded in giving area employers a run for their money. Five times between 1890 and 1920 black and white union miners overcame formidable obstacles to mount district-wide strikes against the South's most powerful industrial employers. Twice in the first two decades of the twentieth century state officials declared martial law in the Birmingham district and dispatched troops to put down strikes approaching insurrectionary levels. If the record of interracial unionism in the Birmingham district is exceptional, it is nevertheless instructive for what it reveals about the early twentieth-century South. Far from being indifferent to or repelled by the excesses of Jim Crow, white employers were the system's chief beneficiaries. White workers, however, were among its victims. In their attempts to mount an effective challenge to the power of their employers, black and white Alabama miners advanced a vision that—while falling short of thoroughgoing racial egalitarianism—nevertheless had to be snuffed out lest it become infectious among others at the bottom of southern society.

Prologue: Judgment Day, 1908

"It was a third of a century ago," Frank Evans reminded readers of the *Birmingham Age-Herald* in August 1908,

> that the people of Alabama by rigid force . . . stopped the advance of a threatening peril which endangered our social fabric[—]the inculcation in the minds of blacks [of] the idea of social equality. [This] terrible poison . . . was . . . applied for political purpose by carpetbaggers from the north, and for a time the cloud seemed ominous, but the Caucasian blood of this state was aroused to resentment and to the defense of the home fireside.
>
> When today this correspondent saw the comingling [*sic*] of whites and blacks at Dora, where he beheld the sympathetic arms of a negro . . . embrace a white speaker to impart to him a secret of his bosom, in the very presence of gentle white women and innocent little girls, I thought to myself: has it again come to this?
>
> One interesting feature of this meeting . . . was the singing of a hymn by the assembly. As [UMW President] Fairley took his seat a "square note" music teacher led the sacred warning hymn, "Are You Ready for the Judgment Day," and white and black, male and female joined in concert.
>
> Just what bearing the musical introduction of this all-important question has upon the present status of affairs at Dora I do not fully understand, but I thought as I looked upon the assembly of idle men, heard of dynamite fury and beheld the presence of armed officers, civil and military, and saw the glittering of handcuffs here and there, that the question is not inappropos at this time.[1]

Measured against other men of his race and social standing in turn-of-the-century Alabama, Frank Evans was not a particularly rabid race-baiter. Nor was he normally inclined to public grieving over the Lost Cause, the dark, bygone days of Reconstruction, or the tragedy of black Republican rule. Less than a decade earlier, in fact, Evans had "openly welcomed" black support for his mayoral bid, and the desertion of prominent Birmingham "race leaders" to the Democratic Party ticket had been key to his electoral success. Evans's peers in Birmingham's business elite would have scoffed at any suggestion that his was a reactionary, backward-looking creed; like him, they considered themselves "progressives," harbingers of a new, enlightened social order in the region. But the odious dispatches he penned to the Birmingham press in the fall of 1908—for which Evans was handsomely rewarded by district coal operators—reflected the state of panic that he and Birmingham's white elite had been thrown into by the remarkable insurrection gathering force before them.[2]

On July 6, 1908, four thousand black and white Alabama coal miners had gone on strike after commercial operators in the Birmingham district demanded a 17 percent reduction in wages. The strike was an all-or-nothing affair for District 20 of the United Mine Workers (UMW), a last-ditch attempt to hold on to some semblance of organization in the Alabama coalfields after a half-decade of debilitating, increasingly defensive skirmishes with local operators. Formed out of the remnants of a statewide miners' organization in 1898, Alabama's District 20 had never enjoyed any real stability in the Birmingham district coalfields. From a peak of some ten thousand members just after the turn of the century, the union found its very existence challenged from the outset by local operators. The district's largest employer, the Tennessee Coal, Iron, and Railroad Company (TCI), had balked at renewing its contract with the union in 1903, and in May 1904 it led the other major captive mine operators,[3] including Sloss-Sheffield and the Republic Iron and Steel Company, in announcing their refusal to sign a contract and their determination to operate the mines on an "open-shop" (nonunion) basis. These three companies alone employed nearly half of all miners in the Birmingham district, and the UMW had little choice but to respond with a district-wide strike.

That strike lasted over two years, exacting a heavy toll on the UMW's power in the Birmingham district and sapping the morale of its members. Relying heavily on convict labor to guarantee a steady supply of coal, the mine owners managed to maintain their operations without any serious disruption in output. At many mines the notorious "contracting system"[4]—through which operators contracted with individual foremen for the labor of predominantly black work gangs under their control—was successfully reintroduced after having been discontinued under union contract. The major companies forced their employees to sign "yellow dog" contracts, which stipulated that they would not join unions. By the time the strike ended in August 1906, UMW influence was confined to some thirty or forty of the small, commercial companies, covering less than a quarter of all miners in the district.

The business depression that followed the financial panic of 1907 reinforced the inclination of those operators still holding contracts with the UMW to follow the example set by the captive operators. Their July 1908 demand for a sharp wage reduction was understood by the UMW as a final effort to dislodge the union from Alabama. Both sides geared up for a major confrontation. Faced with extinction, District 20 called a strike on July 6 and launched a major effort to reorganize the coalfields, dispatching organizers across the district and attempting to revive the union in mines where it had been dormant for years. On July 12, twenty-three of the major coal companies published a statement in the local press declaring their "adherence to 'open-shop' principles" and vowing to rid the district of coal unionism.[5]

The UMW recognized early on that the union's very existence was at stake in the confrontation. To prevail they would not only have to bring their existing membership out on strike but also make substantial inroads at those mines that had been conducting themselves on an open-shop basis, including especially those run by the major captive operators. The UMW's early success must have stunned the employers: thirty union locals were organized within the first week and some seven thousand miners joined the four thousand already out on strike. By the end of July District 20 claimed twelve thousand members, and they had planted UMW organization at a number of nonunion mines. Their success in building a local among the five hundred black miners at Blue Creek, who only several years earlier had been imported to the district as strikebreakers, illustrates the ready reception that UMW organizers met across the district.[6]

Though mine operators refrained from publicly acknowledging the strike's effectiveness, their panicked response reflected deep bewilderment at the turn of events. They organized a major effort to import strikebreakers into the district, dispatching labor agents as far away as Ellis Island to bring relief. At the larger mines, operators constructed stockades around their operations. At Mary Lee mines, the *Age-Herald* reported, "a space ha[d] been cleared" that was "brilliantly lighted" at night, "making it almost impossible for anyone to get close to the place without the guards seeing them." As the strike wore on, however, it became unclear whether such precautions were being taken to fend off intrusions by "agitators" or to prevent imported strikebreakers from deserting. When authorities visited the Banner mines to investigate charges that men were being forced to work against their will, none of those queried replied that they were, "though 15 decided to leave."[7]

From the beginning of the strike, heavily armed "deputies" were deployed throughout the district, and by the end of the month some sixty "Texas sharp-shooters" had been imported from the west and deputized by the companies. Despite these measures, however, the operators' efforts to replace striking miners in the first few weeks were frustrated by the strikers' impressive militancy and organization. A *New York Times* correspondent lauded the miners' remarkable ability "to anticipate arrivals of strikebreakers hours ahead of time," attributing it to "their excellent detective system." Frequently, crowds of up to a thousand striking miners rallied at the rail depots to "welcome" incoming strikebreakers, appealing to them for support. "Get off and join us and we'll feed you and give you five dollars a week!" a gathering of black strikers yelled out to strikebreakers at the Adamsville rail depot. "We were out there day and night," one striker later recalled, asking the people "did they know where they was going? Well, they said they was going on a farm [but] we told them no you are not . . . you are going over here to a coal mine to break a strike."[8]

As company-paid deputies and later soldiers intervened to prevent such interactions, union miners resorted increasingly to violence to stop the influx of scabs. Reports from a range of sources agree that their efforts were effective. By mid-July, the *New York Times* reported that "at least a half dozen lives had been lost, and the well-armed, well-fed, and well-housed strikers are in command of the situation." Following a fierce gun-battle near Brookside on July 17, described by one newspaper as "the most exciting battle . . . since the Civil War," a "train load of thirty deputies escaped through a tunnel in the mountain," leaving the area "in the hands of nearly a thousand sympathizers of the striking miners." At Pratt City "hundreds of strikers gathered about the convict mines . . . and threatened to turn loose the convicts as they were being conveyed back to prison." From Wylam came a report that the air shaft at the number 5 mine had been blown up. Strikers dynamited the houses of nonunion miners throughout the district, and at Pratt City they targeted the home of the TCI mine boss Thomas Duggan. In early August, twenty-seven people were arrested in connection with an attack on a train at Blocton, among them "8 negro men, 1 negro woman, and the balance Slavs," most of whom had been strikebreakers themselves during an earlier strike. Surveying the situation developing in the coalfields, one prominent Birmingham citizen insisted that events had overstepped the boundaries of a mere strike: "It is simply a revolution!" he declared indignantly.[9]

As the strike's effectiveness began to impress itself upon the operators, their penchant for vigilante methods began to dominate public discussion about resolving the dispute. Backed by powerful allies in Birmingham's business community, operators called on Governor Braxton Bragg Comer, himself an owner of the state's largest textile mills at Avondale, to intervene with troops. "If the state has no law to deal with the strike the citizens should make one," the real estate magnate Robert Jemison asserted. The owner of the Shelby Iron Company, J. E. Shelby, concurred, prodding an audience at the city's Commercial Club with the suggestion that "If the governor is not going to take steps to declare martial law, the public ought to get together and let him know what they want." The industrialist J. W. Bush went further, promising that he could "gather a force of Confederate veterans who could disarm the strikers if only given the power to do so," and from south Alabama came the taunt that "If [UMW President] Fairley and his black co-conspirators had invaded [here] nothing further would be needed but the coroner." In the end, however, no such measures would be necessary: Comer obliged the business community with troops and a declaration of martial law.[10]

From the beginning, public discussion of the events had focused prominently upon the union's racial composition. One of the most troubling aspects of the strike from the perspective of the operators and their allies in the business community was the sharp challenge it posed to the area's rigid racial order. The operators Walter Moore and Guy Johnson penned a letter to the *Birmingham Age-Herald* characterizing the UMW's interracial policy as "a direct insult to our southern traditions." Frank Evans, the former mayor of Birmingham, under contract from the Alabama Coal Operators' Association (ACOA), fed his readership a steady diet of racial hysteria, complaining that "the chief white leaders of this strife" had "fired the minds of ignorant and vicious blacks with the statement that under the contract system the negro was doing all the work and the white man getting the pay." Elsewhere he protested the remarks of a "negro speaker" who had stirred up black miners by telling them that "they were as good as white men, and should demand high places of honor and trust." The editor of the society page, Dolly Dalrymple (whose "specialty" it was to "put witty sayings into the mouths of darkies"), transformed her gossip-filled column into a diatribe against the UMW. Horrified by the union's attempt to organize women's auxiliaries among miners' wives, Dalrymple mounted an aggressive rearguard defense of the color line: "White women and black women meeting on the basis of 'social equality' indeed! White men holding umbrellas over negro speakers! Black men addressing white men as 'brother'! The women of our fair southland resent it."[11]

Daily reports from the coalfields added to the perception, widely embraced by Birmingham elites, that their social order and the racial hierarchy it embodied was threatened with imminent collapse. Evidence mounted that the strike had punctured the aura of invincibility that enshrined white supremacy in the district. In spite of the racial conservatism of its leadership and many white rank-and-file miners, the UMW's militancy had conferred a degree of legitimacy upon black resistance that it never enjoyed in more placid times. The strike had upended racial protocol and with it the ritual deference that allowed white paternalists to convince themselves that theirs was a harmonious society. Black strikers were "armed to the teeth and . . . everywhere in predominance," warned Major G. B. Seals of the Alabama Guard, and press accounts left little doubt that the weapons were for practical rather than decorative purposes. When deputies challenged two armed blacks at Blue Creek in mid-July and "told them to stop parading the roads with their weapons," the strikers "stood up and gave the deputies a fight, firing at them with considerable precision." One striker was killed in the battle, "literally punctured with shot." Another serious gun battle was narrowly averted at the Adamsville rail depot, where deputies attempted to disperse a crowd of seven hundred miners who had gathered to "attempt to entice [strikebreakers] to join their ranks." "Twice a negro in the crowd raised his gun, preparing to fire," a reporter recounted, and "as many times the deputies drew a bead on him . . . but the snapping of a trigger would have resulted in a fight." Of seventeen miners arrested for dynamiting houses at Acton, thirteen were black. Two black men and a black woman were the only strikers arrested for setting the charge to the TCI foreman's house at Pratt City. Clearly black strikers suffered the brunt of repression in the coalfields; but by all accounts they were also the most determined strikers.[12]

Black unionists did not hesitate to deal harshly with strikebreakers of either race. A white strikebreaker at Pratt mines was startled on his way to work in mid-August when two heavily armed black strikers "step[ped] out from behind some bushes in front of him." Racial solidarity may have reinforced whatever sympathies black unionists felt for those packed onto the trains arriving daily from the Cotton Belt, but it seemed to offer little protection to those who had willfully defied the UMW's strike order. At Short Creek, a black union stalwart "entered the mine through a private, unused narrow passage . . . and told the negro miners that they had better cease work immediately . . . that the strikers had sworn vengeance upon all who continued to work." When a black miner at the Banner mines declared in early August his determination to break ranks and return to work, Ned Harris, a black UMW member, "cursed at this, and replied that any man who would do that ought to be killed . . . then pulled a pistol and shot [him]."[13]

For operators accustomed to extolling black loyalty, the specter of a social challenge that overstepped the bounds of industrial conflict seemed a prospect too appalling to tolerate, and from early August onward the race issue figured prominently in their attempts to win public support for suppressing the miners. Operators calculated that their best chances for undermining the strike lay in driving a wedge between the "ignorant whites and blacks who have been idling away these many weeks" and the "better class of white union miners," who they supposed "would return to work at once but are afraid of assassination."[14]

Dividing the union along racial lines was more difficult than its adversaries might have imagined, however. If black strikers' militancy was a source of deep despair for Birmingham elites, they could take little comfort from the union's flagrant disregard for racial protocol. Though it never lived up to the lurid egalitarianism ascribed to it by its enemies, the UMW's interracialism represented an implicit challenge to the racial status quo, and the bonds being forged between black and white miners threatened to permanently derail Birmingham from its path to industrial prosperity. The willingness of black and white strikers to join forces—frequently in arms—against the operators seemed to many to signal an unwelcome blurring of racial lines. The spectacle witnessed by one correspondent at Jasper, where "a brass band led a parade through the streets" in which "negroes as well as whites bore red flags, and black men were among the principal speakers," seemed to Birmingham's men of wealth to portend certain disaster.[15]

The operators' turn to the race card coincided with a steady escalation of violence in the coalfields. At Brighton in early August two company deputies were charged with murder after they snatched the black union militant William Millin from jail and lynched him. Just over a week later the union miner Jake Burros was hanged from a tree after being arrested for allegedly dynamiting the house of a strikebreaker. With tensions mounting, a delegation of prominent Birmingham citizens threatened the UMW vice president John P. White that unless the strike was ended they would "make Springfield, Illinois [where a vicious race riot had taken place only weeks earlier] look like six cents." Governor Comer became increasingly adamant that the strike must be defeated, dispatching troops to the strike zone and concentrating his remarks on the danger it posed to white supremacy.[16]

With the threat of racial conflagration looming over them, the UMW leadership began to fold under the pressure. Comer summoned UMW officials to his office on August 25 and informed them that "members of the legislature in every sector of Alabama" were very much "outraged at the attempts to establish social equality between black and white miners." Declaring that he would not tolerate "eight or nine thousand idle niggers [sic] in the State

of Alabama," Comer threatened to convene a special session of the state legislature unless officials ended the strike immediately. In a maneuver that revealed the efficacy of the operators' race-baiting campaign and its own accommodation to racial conservatism, the UMW attempted to deflect the
"social equality" charge rather than confront it directly. At one point the
national UMW offered to "transport out of the state every Negro who was
on strike and make it a white man's fight," but to no avail. Invoking public
health concerns as a pretext for a final assault on the union, Comer ordered
the military on August 26 to cut down the tents erected by striking miners
throughout the district.[17]

Four days later, union officials declared the strike over. Judgment Day had
arrived, though decidedly not in the form that operators had dreaded early
in the strike. The UMW had suffered a colossal defeat in Alabama, a debacle
that sealed the fate not only of industrial unionism in the coalfields but of
any possibility that black workers might extend their challenge to the racial
status quo. In attempting to explain the defeat some weeks later, John P. White
attributed the loss to the deliberate manipulation of racial prejudice. "Finding themselves defeated on the industrial battlefield," he claimed, operators
"resorted to drastic methods . . . the columns of the press began to teem with
appeals to [prejudice], and they injected the race question. . . . Any one who
is a student of Dixie Land knows what effect this has. . . . The entire South is
slumbering with racial hatred, ready to break out in widespread racial conflagration. . . . Yet it was necessary in order that the operators would not be
defeated. We find ourselves in a very precarious condition." The truth of
White's somber assessment would be revealed in the weeks and months
ahead. The UMW was shattered beyond recovery, at least in the here and now.
For as far into the future as anyone could see, Alabama operators would have
things their way.[18]

1. The Operators' Dilemma

DISTRICT OPERATORS HAD GOOD REASON to savor their newly bolstered supremacy in the aftermath of the 1908 strike. Prospects for extended industrial peace seemed bright, clearing the way for capital improvements that would have been shelved as unnecessary risks while the UMW remained viable. The employers now enjoyed unilateral authority in regulating wages and setting the conditions under which miners would labor. Even so, their celebration was to be short-lived. In attempting to rally public support for a protracted campaign against the miners' union, Birmingham district operators had invariably identified the UMW as the chief threat to regional prosperity. The eradication of UMW power ranked so high among the operators' priorities that any casual observer of the Birmingham scene might reasonably conclude that coal unionism remained the only block on industrial growth. But fervent antiunionism had combined with a zealously optimistic booster mentality to impede an honest assessment of Birmingham's industrial prospects.

So long as the UMW provided a foil for the district's shortcomings, mine owners could delay coming to terms with a host of equally intractable problems related to labor efficiency and supply. Rather than delivering relief, however, their victory over the union ushered in a day of reckoning. Like their counterparts elsewhere throughout the industrializing South, district coal operators finally confronted the crux of the region's labor problem: the difficulty of securing and holding an adequate supply of labor under wages and conditions that free workers might reasonably spurn. In ways that they had failed to anticipate, victory over the UMW had merely exacerbated the operators' dilemma.

The most urgent problem confronting district coal operators was the intense bitterness that their own actions had instilled among the mining workforce. The heavy-handed deployment of troops throughout the mineral district, the business community's panicked resort to vigilantism, the provocative injection of the race issue, and Comer's final, humiliating directive to cut down strikers' tents had raised tensions in the coalfields to the brink of open insurrection. When UMW officials announced the termination of the strike on September 1, their words did little to lift that tension and in some areas were almost completely ignored by rank-and-file miners. Of the four hundred "proud, firm and determined" union delegates who assembled in Birmingham to ratify the back-to-work order on September 3, many had been sent with explicit instructions from local unions to defy their national leadership. Large numbers of miners clustered in the streets outside Maccabees' Hall to "make their intention of not returning to work known to the union leaders," while delegates inside fiercely debated whether to continue the strike without the sanction of the national UMW. In the end, a majority of delegates consented to ending the strike, but few did so without harboring the hope that strike action could be renewed in the near future. The UMW vice president John P. White, who had been dispatched to Birmingham to direct the strike, recalled the tallying of the final vote as "the saddest moment of [his] life," when "brave men wept like children, fully conscious of our supremacy on the industrial field, and yet hopeless to prevent victory from being taken from us."[1]

Against the backdrop of such deep divisions, union leaders encountered bitter resistance in implementing the return-to-work order. More than a week after the strike had been officially ended, the *Birmingham Ledger* reported that the union commissary at Wylam was still feeding three hundred people a day. Miners there—who had been among the last to respond to the strike call—held a mass meeting on September 11 to express their objection to the unilateral ending of the strike, now widely condemned as a "sell-out." Governor Comer ordered troops to remain in the coal district after the strike's official termination to spurn any efforts by dissident locals to continue independently, and holdouts submitted letters to the labor press urging miners to prepare for a new round of confrontation. "Pay your poll tax and stick to your union," advised R. A. Statham from Lehigh, "and as far as buying a gun every miner should have one." "Every man in the district is ready to strike again," J. T. Durrett assured readers of the *Labor Advocate*, "and is ready and waiting for the summons to come." From Searles a UMW stalwart reported that "hundreds" of miners in the southern portion of the district were "turning to socialism like fish in a pond that's been dynamited," because it offered "remedies for the very things that caused all of this intense suffering and trouble by the government."[2]

Many miners harbored a deep antipathy toward the coal operators and an eagerness to reengage them in a fight to the finish. But it became increasingly clear in the weeks following the strike that while a substantial minority might welcome the chance for renewed industrial action, there was little chance that a new effort could alter the overall outcome. Despite their remarkable success in shutting down coal production, union miners had been singularly frustrated in their attempts to deflect the fierce campaign waged against them in the court of public opinion, and the possibilities for an effective renewal of the strike became more remote with each passing day. The flash of anger and militancy evident in the days after the strike's termination gave way to a palpable sense of demoralization among miners and, conversely, to provocative flaunting of their newly won hegemony on the part of area operators.

Rearguard, localized defiance could not diminish the scale of the defeat: the strike's outcome had fundamentally altered the balance of power and settled for the foreseeable future the question of who would rule the coalfields. The confidence with which operators dismissed the threat of a new strike can be gleaned from their brazen attempts to reassert management prerogative in the wake of the defeat. Across the district, operators forced returning miners to sign housing leases that stipulated "what roads the lessee shall travel and who shall visit the lessee, and whose company he shall entertain" and inscribed their right to evict miners from company housing on a single day's notice. The ACOA, founded in the final days of the strike as a command center for the operators' campaign,[3] appointed a committee to "regulate the movement of men from camp to camp," hopeful that they could end internecine competition for skilled labor and hold the line on wages by restricting miners' ability to move about the district.

The operators' formidable repressive apparatus, which had played such a conspicuous role during the strike, was reorganized under the auspices of the ACOA. Operators contracted with the Austin-Robbins detective agency immediately after the end of the strike to "keep a close watch and try to prevent further organization of unions." For several months after the strike its duties included keeping track of "the number of loyal [union] men in different camps," but by early 1909 operators acknowledged that since it was "not known that any organizers were [left] in the district," there was "really very little work to be done." Finally, the ACOA attempted to seal its rout of the UMW by dislodging the union from the handful of mines where it had managed to hold on to some semblance of organization. A special committee of three ACOA members was appointed to visit those few small, independent operators still holding contracts with the union "to induce them to operate open shop mines." Their ability to carry out such provocations without incurring even token industrial action attests to the demoralization rampant among rank-and-file miners.[4]

As the new order took root in Birmingham district coalfields, experienced miners began to acknowledge, through their actions if not their words, the scale of the defeat. Lacking realistic short-term prospects for reversing the setback in Alabama, skilled miners—disproportionately white—took the path of least resistance and began emigrating north in large numbers. In an effort aimed at salvaging what remained of the union's reputation among district miners and imposing a labor quarantine upon Birmingham district operators, the UMW offered to transport miners out of the state to the unionized northern fields. Hundreds took the opportunity to put the strike—and Alabama—behind them. For those who had been "blackballed" for union activity emigration offered the best and sometimes only hope of eking out a living. For others the lure of higher wages and a fresh start was ample incentive to pack their belongings. The *Labor Advocate* predicted that the "best miners" in the state would "refuse to submit to these injustices" and leave for "states where the wages are from $1 to $2 per day more than here." By late September the ACOA expressed its own concern that the exodus of skilled men out of the district was taking its toll on the labor supply, a complaint that won them little sympathy from organized labor. The "mines will be worked by the riff-raff shipped in to take [the strikers'] places," *Advocate* editors reasoned, with some satisfaction. "[The operators] have sown the wind and will reap the whirlwind."[5]

The retribution wished upon the coal operators by the *Advocate* carried heavier consequences than either they or the operators at first comprehended, and its realization over the next several years underscored the contradictory results of the 1908 strike. The TCI president Don Bacon, who set district operators on a collision course with the UMW in 1904, had considered the elimination of union influence an essential condition for restoring long-term profitability. The "conditions forced upon management by the union had become intolerable," Bacon complained in 1905, and unless company officials succeeded in reestablishing "control . . . over the business of the company," he told the *Wall Street Journal*, "all hope of permanent, successful competition with the products of other districts must be abandoned."[6]

With their rout over the union in 1908, mine owners reclaimed the authority Bacon deemed so essential. In their 1909 annual report, ACOA officials boasted that the Birmingham district coalfields had been "singularly free from labor troubles" since the end of the strike, and similar results were recorded nearly every year until the middle of the following decade. Their own confidential investigation reported during the same year that the once-formidable UMW had nearly vanished from the Alabama fields: only 278 miners remained in the union from a membership of nearly twenty thousand at the peak of the 1908 strike. Federal officials in 1910 reported that the Alabama mines were "peculiarly under the control of the employers," and the labor press acknowledged that the defeat of the UMW had left labor relations in the district coal mines "a one-sided affair. [The operators] are organized and labor is not." An investigation into conditions in the Birmingham district four years after the strike corroborated these accounts. "The operators have had things their own way," the labor reporter John Fitch observed in 1912, and their authority was "exercised with a high hand by every company in the district." A UMW delegation sent to the Alabama fields in 1913 to survey prospects for reviving the union found an atmosphere of "fear bordering on terror," where "if it becomes known that [miners] have met or talked with any of our representatives they are immediately discharged." By all accounts the strike was a pivotal event in the operators' efforts to assert their control.[7]

The singular irony of the strike settlement, however, is that the operators' success in driving out the UMW exacerbated an even more daunting problem: the district's chronic "shortage" of skilled, disciplined industrial labor. Ideally, of course, area employers preferred to have at their disposal a large labor reserve that embodied in equal measure the qualities of skill and tractability. But insofar as their real human material at hand fell short of this ideal, some of the region's most avid promoters were unconvinced, prior to the strike, that driving the union out of the coalfields was the most pressing order of business. In some quarters, where such a campaign would make it more difficult to attract and hold the "higher order of workmen" to the district, it met with surprising ambivalence. "It would be better to have the entire Birmingham district unionized," one area booster advised heretically in the pages of the *Manufacturers' Record* shortly before the strike, "provided that they [sic] would secure an adequate supply of labor, than to attempt to keep on with the miserably inadequate supply now at hand." As the drift of skilled miners out of the district threatened to become a stampede, operators might have wondered if the costs of their success had been too high.[8]

The UMW seems to have discerned the pyrrhic quality of the operators' victory within three weeks of having issued its back-to-work order. The *UMW Journal* pointed out that although their intransigence had succeeded in forcing a return to work, Alabama operators had acquired public notoriety that was "not good advertising for employers of labor who are anxious to get a better class of workmen." Self-respecting miners would leave the district in droves, the UMW anticipated, "unless they are natives who have always been handled that way." While the union's predictions were obviously laced with deep bitterness and shaped by its interest in encouraging such an exodus, evidence from all quarters confirms that the operators had painted themselves into a difficult corner. J. T. Durrett, a union miner, reported within two months of the strike's conclusion that the men who had been brought in to break the strike at Searles were bankrupting the operators by their inefficiency. "All I want is for the union to remain dead in Alabama about twelve months longer," he quipped sarcastically, "and the company will *make* the men organize."[9]

Operators responded to the immediate crisis with a series of stopgap emergency measures aimed at stemming the outflow of labor from the district. The drain on mine labor had become so acute by the following autumn that the ACOA organized its own Labor Bureau, sending the TCI official Major F. B. Dodge[10] into Mississippi's "boll weevil district" to recruit black laborers. Individual operators dispatched labor agents as far as New Orleans in an attempt to bring new labor into the district. But recruitment efforts would be worthless unless accompanied by a systematic plan for preventing those miners already resident from leaving the district. Mine owners therefore supplemented their attempts to secure new labor with restrictions against the mobility of experienced miners, and for obvious reasons it was black miners who bore the brunt of these measures.

The elaborate web of legislation that downstate planters had fashioned to limit the movement of black agricultural labor provided operators with a valuable lever in their attempts to fasten mine labor to the coal district.[11] Laws punishing vagrancy and debt were applied energetically throughout the mineral district, frequently supplemented by more extraordinary, extralegal measures. A "diplomatic letter" forwarded to local railroad agents by the secretary of the ACOA foreshadowed the employers' response to large-scale emigration during the war. In it, operators requested that the railways "refrain from running negro excursions out of [the] district" on the grounds that they "would have a tendency to disrupt the working force and in many cases carry labor out of the district permanently."[12]

The mine owners' attempts to recruit and hold skilled labor were frequently undermined by their vigilance against resurrection of the UMW. Poststrike conditions provided for a "peculiar situation," reported the *Labor Advocate*, wherein "men are wanted and they are not wanted." The tribulations endured by A. J. Gibson, a white skilled miner, while trying to secure work at the Short Creek mine in early 1911 illustrate the operators' difficulty in balancing vigilance with the recruitment of competent, skilled labor. When Gibson presented himself, neatly dressed, to the mine foreman at Short Creek, his appearance seemed so exceptional to company officials that, after initially offering him a job, management decided in a nighttime meeting "that they could not employ him from the fact that he looked too good," suspecting him of being a union organizer. Officials were so wary of Gibson's presence that they dispatched a company deputy to eject him from his nearby hotel room. "X-Miner," a witness to the affair, concluded that "while [the operators] want men, they want to see a lot of those miners come with bundles on their backs and trunks under their arms—those who can live on 25 cents a day."[13]

To editors at the *Labor Advocate,* it seemed that only those miners willing to live in company housing and spend "all, or nearly all" of their earnings in the company store were being kept on. Certainly mine operators devoted considerable attention to purging their camps of malcontents who might be vulnerable to renewed labor agitation. At one camp, where the pit boss seemed to be "putting in all of his time trying to discern the good from the bad," miners reckoned that "if he succeeds in his undertaking there won't be a man left . . . in six months." Likewise, as the *Advocate* insisted, company housing and the camp commissaries were important sources of profit. But the suggestion that short-term profiteering overwhelmed the operators' interest in broad development of their labor resources seems naive. As their cohesiveness during the strike had shown, most medium-to-large-scale mining operations could see beyond their immediate interests and were willing to subordinate short-term gain to the search for durable, long-term solutions to their labor difficulties. Their motivations were more complex than the *Advocate*'s interpretation allows.[14]

The contradictory features of the labor strategy pursued after 1908 suggest that the employers were being pulled in different directions by potent though conflicting imperatives. The urgent need to develop and hold a reasonably skilled workforce led the largest operators to the conclusion that systematic reform was necessary. At the same time, equally compelling pressures to hold labor costs to a bare minimum—felt more acutely, perhaps, by smaller, independent operators—imposed definite limits on these ambitions. Their frustrating attempts to find a way out of this dilemma over the next several years would lead operators to embark on a comprehensive overhaul of labor relations in the Birmingham district and establish a new industrial order that bore this incongruity at its heart. The new order would fuse one of the most ambitious applications of "progressive" company welfare in the United States with a system of racial paternalism rooted in the traditions of the Old South. Not only Birmingham but the district as a whole would come to resemble an "iron plantation" where "all the pestilent inheritances" of slavery were combined with "the worst excesses familiar to Northern industry."[15]

In their various attempts to resolve the persistent labor difficulties confronting them, Birmingham's steel and iron masters found themselves trapped in the same labyrinthine predicament that beset industrial employers elsewhere throughout the New South. Many of them acknowledged the need to modernize, not only through building up the industrial infrastructure of the region and deploying new technology in their operations but by recruiting skilled workers to the district and, where necessary, bending local custom to accommodate a more cosmopolitan mix of peoples and ideas. At the same time, however, certain elements of the old order, including especially the myriad advantages offered them by the racial subordination of black workers under Jim Crow, overwhelmed any urge to contemplate a decisive break with traditional southern labor relations. White supremacy won a new lease on life in the New South's industrial showcase. But the circumstances that breathed new life into antebellum custom, which guaranteed continuity between the old order and the new and paradoxically imposed the most debilitating constraints on development in the Birmingham district, were beyond the immediate control of local industrialists.

The most important of these was the difficult competitive position that northern domination of the steel industry had carved out for the district. By the turn of the century Birmingham had earned for itself a reputation as the "city of perpetual promise," a moniker that acknowledged the district's vast industrial potential as well as its invariable tendency to fall short of expectations. With regard to the area's material resources, friend and foe seemed to agree that Birmingham was destined to develop into a major industrial power. "I will not say that Birmingham will furnish the world with iron," the English ironmaster Lothian Bell had remarked after his tour of the district in 1883, "but I will say she will eventually dictate to the world what the price of iron shall be." Six years later Andrew Carnegie declared the district "Pennsylvania's most formidable enemy," and the close proximity of all the elements of steelmaking struck more than one booster as "an advantage that can not be overcome." Wielding an apt metaphor for a city acquiring a reputation as "Murder Capital of the Nation," Herbert N. Casson boasted in *Munsey Magazine* that from one of Birmingham's furnaces "it would be quite possible for a sharpshooter to stand on the water-tower and with a rifle send a bullet into the mines out of which the ore was dug. With a revolver he could put the men in the limestone quarries in danger; and with a peashooter he could annoy the workers at the coke ovens."[16]

By the early twentieth century, Birmingham had experience enough with industrial downturns to take the edge off of the unguarded optimism of its early years. Even so, the acquisition of the district's leading industrial enterprise, the Tennessee Coal, Iron, and Railroad Company, by U.S. Steel in 1907 set off a new round of speculative euphoria. Editors at the *Age-Herald* predicted that the new owners of TCI would "make the district hum as it has never hummed before," and the *Birmingham News* predicted that "superiority of product and cheapness of manufacture will conspire soon to make the Birmingham district the largest steel producing center in the universe." Even the labor press welcomed the entry of the steel giant, hopeful that the departure of the TCI president Don Bacon, dubbed the "Duke of Ensley" by organized labor after a protracted series of bitter confrontations, would usher in an easier time in the years ahead.[17]

Practically alone in his dissent from the elation pervading the district, the Tennessee industrialist Hugh W. Sanford worried that the entry of the "Steel Trust" would arrest the development of the southern steel industry rather than advance it. Sanford agreed that there was "no reason in the world" why steel could not be manufactured in the South "at the same cost advantage" enjoyed in pig-iron production but warned that any advantages attending U.S. Steel's entry would be offset by a pricing system designed to safeguard Pittsburgh's supremacy. By selling steel "uniformly at the Pittsburgh price . . . plus freight rates from Pittsburgh regardless of where the shipments originated from, competition with any other part of the country" would be "forestalled." Pittsburgh was to be "Lord and Master," Sanford warned, and the "Pittsburgh Plus" pricing differential would have a "boa-constrictor effect" on the southern steel industry.[18]

"Pittsburgh Plus" was not alone responsible for the difficult position that Birmingham iron and steel found itself in during the early years of the twentieth century.[19] The giddy optimism exuded by district boosters, aimed partially at attracting outside capital, had consistently overstated the region's material advantages. The pig-iron industry, mostly unaffected by U.S. Steel's pricing policy, displayed a similar inability to break from a pattern of fitful, lackluster growth. Metallurgical problems posed by Birmingham's low-grade iron ores, the absence of home markets in the South, and the prohibitive costs of shipping coal and steel and iron products to manufacturing centers in the North all reinforced whatever handicaps had been introduced by southern subordination to Pittsburgh.[20] But the cumulative effect of these disadvantages was unmistakable. In the long term they reinforced Birmingham's preoccupation with the one resource that afforded it a clear competitive advantage over steel producers elsewhere: a seemingly limitless supply of cheap labor.

In acknowledging native labor's strategic importance to industrial advance, Birmingham employers articulated a local variation of the regnant New South gospel of progress. The most striking contradiction inscribed in that formula was its candid acceptance that the continued impoverishment and low living standards of ordinary southerners would serve as the cornerstone of regional prosperity. The apostles of the new order never failed to boast of the region's tremendous natural potential. "Our coal supply is exhaustless," Henry Grady had reminded a Texas audience during a typical oration in 1887. "In marble and granite we have no rivals," and "in lumber our riches are even vaster," he insisted. "Surely the basis of the south's wealth and power is laid by the hand of Almighty God."[21] But in their attempts to lure northern capital to the region, Grady and his disciples rarely neglected to emphasize the South's favorable labor market as its most stunning asset. Dismissing the region's natural advantages as being of "minor importance," a group of New England industrialists identified lower labor costs, which they estimated at 40 percent below those in the North, a longer working day (24 percent longer in North Carolina than in Massachusetts), and the apparent lack of a "disposition to organize labor unions" as the chief advantages of operating in the South.[22]

As the premiere industrial center of the New South, the Birmingham district embraced this peculiar conception of progress more enthusiastically than its southern neighbors. Wages in the Birmingham steel industry were consistently below national standards. While in 1911 steel laborers averaged seventeen cents an hour in Pittsburgh and Chicago, Birmingham employers divided unskilled labor into two grades, paying thirteen and fifteen cents respectively. And although area steel makers insisted that rates were not racially based, observers noted that it was a "general truth" that "white laborers earn[ed] a little higher rate than Negroes." The desperation driving a steady stream of black Alabamians north out of the Black Belt guaranteed employers a steady supply of reserve labor willing to work at low wages. Throughout the first two decades of the twentieth century the state's agricultural laborers registered the lowest pay in the nation, at less than a third of the national average. Their counterparts in neighboring Mississippi, another important source of industrial labor for the district, ranked second to the bottom. And wages alone told only part of the story; Alabama's industrial laborers worked longer hours than their counterparts anywhere in the United States.[23]

The incongruities at the heart of the New South creed were evident in the coalfields as well. The willingness of the coal operators to brave the near-insurrectionary convulsions that accompanied the 1908 strike was clear evidence that they considered low wages and the open shop not merely preferable to unionization but "absolutely vital" to maintaining their competitive position.[24] One close observer of the national coal industry described the predicament of Alabama operators as a "fight for existence." In spite of the bombast peddled by district boosters, the district's coal seams were "rather pathetic," he wrote. All but one were thin and steeply pitched; the product required extensive cleaning before it could be sold; and the cheapest high-quality coal was held almost exclusively by the large, captive mine operators and therefore excluded from the market. All in all, he estimated, "the cost of production in the Birmingham field [is] vastly higher than . . . in any competing field." Exorbitant shipping rates compounded the operators' problems, with the result that "the distant coal from Tennessee [and the northern fields] sells in the backyard of the Birmingham district."[25]

Birmingham's labor costs therefore figured more prominently as a factor in production than they did elsewhere in the United States. Nationally, wages comprised three-quarters of all production costs throughout the period between 1890 and 1910. But in Alabama, where the incentive for introducing modern technology was permanently blunted not only by geographical conditions but by the availability of a large pool of impoverished laborers, "men, rather than machines, did most of the work." The lowest mechanization rates in the nation reinforced the importance of wages as a factor in the district's competitive position. Over the same twenty-year period, during which the southern states, led by Alabama, increased their share from under 15 percent to 21.8 percent of the national coal market, their labor costs declined conspicuously. By 1909 Alabama could lay claim to both the lowest rate of mechanization and the lowest labor costs anywhere in the country.[26]

Their success in keeping the district viable reflected the operators' impressive ability to parry the workforce's wage demands. At the turn of century Alabama miners worked longer hours for less pay than their counterparts elsewhere in the country. The gap would continue to widen throughout the first two decades of the twentieth century, until by 1922—a year after their decisive defeat in a new round of confrontation—every category of Alabama mine labor rated lower in compensation than the eleven other districts reporting. Moreover, wage inequalities were starker at the lower end of the scale: Alabama miners earned just more than half the hourly wage enjoyed by West Virginia miners in 1922, while drivers, timbermen, and coal loaders in all but four states received more than double the Alabama rate. For good reason, then, the TCI president George Gordon Crawford considered the South "the greatest, best, and cheapest labor market in the United States."[27]

Wage considerations therefore figured prominently in the operators' long-term plans for the district. But their contending requirements for a stable, relatively skilled labor supply dictated a labor policy that seemed at times intrinsically contradictory. The contradiction was most apparent among those mine operators whose coal fed their own steel and iron production, and it is worthwhile to distinguish between their labor requirements and those of the smaller, independent coal operators. The largest mining concerns in the Birmingham district were owned by leading steel and iron companies and functioned as integrated units in operations that included labor intensive coal and ore mines at one end and more heavily capitalized foundries, blast furnaces, and steel mills at the other. Of 105 coal companies operating in Alabama in 1911, the leading six extracted more than half of all coal mined in the state; TCI alone produced nearly 20 percent. While the smaller, independent coal operations generally followed the lead of the larger companies regarding labor policy, the majority of them sold their coal on the commercial market and therefore faced minimal direct competition from them. The divergent requirements of the mammoth corporations and the small, commercial coal operators prefigured possible fractures in the employers' ranks, complicating the search for a coherent labor policy.[28]

While the range of work in the iron foundries and steel plants required a combination of highly skilled and relatively unskilled labor, the coal and iron mines could be operated with a relatively smaller number of experienced miners. Where steel production placed a premium on skill, uninterrupted production figured as the most critical requirement in coal: a steady supply was essential to keep the blast furnaces operating. The smaller, independent commercial operators who sold most of their coal on the local market were excluded from steel production and relatively insulated against national market pressures. They did not necessarily share the steel giants' preoccupation with the dearth of skilled labor or their willingness (or ability) to go to the wall for the open shop. It is therefore no surprise that TCI, which had established its hegemony in the district well before 1908, took the lead in formulating a labor strategy that aimed at holding firm on wages without repelling the skilled labor so vital to the district's future.[29]

The peculiar ambivalence that permeated the operators' disposition toward black mine labor was itself the fruit of this intractable dilemma. Although blacks had mined Alabama coal—as slaves—as early as 1861 and by the early twentieth century constituted a majority of the mine workforce, the operators seem never to have completely reconciled themselves to the permanent employment of blacks in the mines, or at least not in skilled positions. The vulnerability and sheer destitution of refugees from the Black Belt provided operators with an obvious and easily replenished source of cheap labor, and it was inevitable, therefore, that blacks would find employment in some capacity underground. During periods of labor unrest, many operators exhibited a marked preference for black labor, appreciative of the "immunity" from "agitation" afforded them by the easy availability of "cheap docile negro labor." Under normal circumstances, however, the alleged deficiencies that black miners shared with the rest of their race provided employers with a convenient scapegoat for problems large and small.[30]

Like their counterparts in plantation agriculture, Birmingham coal operators embraced a melange of the racial assumptions that had sailed through the tumult of the Civil War and Reconstruction barely touched. A combination of racially filtered "common sense," newly adorned with the pretensions of science, and self-interest guided by more pragmatic, market-driven considerations, these racial assumptions underwent something of a renaissance in the new industrial context, where profitability placed a premium on the efficient deployment of labor. Although blacks were well suited to certain kinds of manual labor (invariably the most grueling and least remunerative physical labor), a number of operators opined, as a race they were inherently unfit for work that required complex thought or mechanical skill. Truman H. Aldrich of the Cahaba Coal Mining Company told a Senate investigation in 1885 that he considered Alabama blacks "the best labor I have ever handled." Llewellyn Johns, a mining engineer for Pratt Coal and Iron, agreed that African Americans made the "best workers" but qualified his compliment by distinguishing between good "workers" and good "miners." Johns's fixation with race provided him with an elaborate formula for structuring the division of labor in Pratt's mines, one that also sorted European nationalities according to their skill, physical strength, and susceptibility to agitation. Though few operators incorporated racial considerations so explicitly in their calculations, many shared Johns's preoccupation with discovering the proper "mix" of native whites, blacks, and immigrants, and race became an essential consideration in the new science of handling labor.[31]

Of all the disadvantages associated with black labor, the most exasperating from the perspective of mine operators was its alleged lack of dependability. They believed blacks to be peculiarly susceptible to "shiftlessness," woefully deficient in matters of thrift, and singularly disposed to "wandering and moving about." "Generally speaking," a reporter for the *Age-Herald* fretted, "the colored worker of Alabama is not a success when he is taken from the cotton field and harnessed to the chariot of coal and iron." The *Survey* reporter John Fitch, echoing the frustrations of local employers, complained that blacks did not seem dependable "even as unskilled labor," of which they constituted over 50 percent in the Birmingham district. "As a rule," he observed, "they appear to be content to do just that number of days' work each week that will afford them a bare subsistence—and not one day more."[32]

Fitch's comments would have received the hearty endorsement of the TCI manager Walker Percy, who had complained in 1903 that of the various ethnic groups working Alabama mines, the "ordinary Negro" worked "the least percentage of time." According to Percy, however, earnings on the little work that blacks could be induced to perform went not toward subsistence but "for three purposes—craps, women, and whiskey. . . . They work for two or three days—just long enough to get money to shoot craps." A Sloss executive expressed his own bewilderment that blacks had a tendency to walk off the job "whenever the notion strikes them." "If there is a show in town, or an excursion on the Fourth of July, or a burial, it makes no difference what the excitement may be," he complained, "they will just drop their work and go off." "Any fantom [sic] of an excuse will induce him to quit," a reporter for the *Age-Herald* concurred. "He will stop work for a revival, for a circus, for his grandmother's birthday." Others complained that blacks brought with them into the mines the rhythm of seasonal labor they had known on Black Belt plantations. Few black miners could be induced to work Saturday afternoons, for instance, and many took extended vacations at the years' end to trek southward where those they had left behind were settling up with planters and celebrating the end of another year's labor.[33]

The inability—or unwillingness—of black laborers to "harness" themselves full-time to the chariot of industry led mine owners to adopt a number of extraordinary measures aimed at minimizing the disruption caused by unsteady labor. Fitch reported that at many of the area mines and mills managers kept "at least fifty percent" more men on their payrolls than were required at any given time "in order to insure a complete force each working day." "We are obliged to surround ourselves with twice as much labor as we need," a Sloss official complained, "because a negro refuses to work more than half of the time." According to Fitch's colleague Shelby Harrison, Alabama operators considered "regularity" one of the chief attractions of the convict system. Where the coal companies employed convicts, he reported, "three hundred men . . . go to sleep at night, and three hundred men get up the next day . . . and this was for 310 days a year." At some mines management attempted to reduce its dependency on black labor by hiring whites wherever possible. "Whenever I can put a white man in the place of a negro, I do," the Sloss-Sheffield president John C. Maben reported. "The only final solution is white labor, and I expect that we shall be driven to bring in Italians and Hungarians from the north."[34]

The possibility that immigration could ease labor shortages and reduce their dependence on black labor seemed to many employers in the district a welcome panacea to their most pressing difficulties, and from the turn of the century onward Birmingham industrialists placed themselves at the head of an ambitious immigration campaign. The most enduring accomplishments credited to TCI's Don Bacon upon his departure from Birmingham were his routing of the UMW during the 1904 and 1906 strikes and, significantly, his importation of "Huns, Slavs, Greeks, Servians, Italians, and Finns to replace . . . unreliable negroes." His successors extended the immigration policy, sending agents north to Philadelphia and New York in an attempt to entice new arrivals southward. Representatives of the district's leading employers provided the nucleus for a local branch of the Southern Immigration and Industrial Association.[35]

Throughout the first decade of the twentieth century Birmingham's business press was littered with hopeful reports from across the South that relief from the unpredictability of black labor was around the corner. On this issue, at least, the interests of downstate planters and Birmingham's "Big Mule" industrialists seemed to converge. Under the headline "Italians Are Better Farmers," the *Age-Herald* reprinted remarks made by the Louisiana planter Colonel F. L. Maxwell to the 1907 Southern Cotton Association meeting in Birmingham. Maxwell's experience on his seven-thousand-acre plantation had convinced him that the Italian farm laborer was "more valuable than the negro." He found them "more thrifty and industrious" than blacks and reported that "many of them are better farmers" but qualified his remarks by identifying two deficiencies associated with Italians. Most bewildering of all was the fact that the typical immigrant seemed "restless and moves around too much." After settling up with the landlord at year's end, Maxwell complained, the Italian "immediately begins to look around to see if there is anything better for him. He visits all the neighboring plantations, and if any offer him anything he will pick up and move." Maxwell's second objection, also significant, was that he could not "talk to [the Italians] and direct them in their work as you can a negro." With a "serious" labor shortage, a large area under cultivation, and substantial holdings of unforested timber, however, Maxwell hoped that he could "get many more of them than I now have."[36]

Captain J. F. Merry, an immigration commissioner for the Illinois Central Railroad, likewise expressed qualified optimism in remarks to the *Age-Herald*'s editors later the same year. While en route to meet a shipload of new arrivals from southern Europe at New Orleans in August, Merry expressed his view that "the negro [was] outliving his usefulness as a field hand." To avert a crisis, he warned, it was imperative that "the agricultural interests of the South . . . make a concerted effort to secure a new labor supply." Though sharing Merry's concern over the gravity of the labor problem and buoyed by signs of an imminent upturn in immigration to the region, the editors tactfully dissented from the commissioner's unflattering assessment of black labor. The problem from their perspective was not that such labor was inferior but that its supply was inadequate. "Negro labor is good as far as it goes, but there are not enough negroes to 'go 'round,'" the paper concluded. "Foreign labor is therefore in increasing demand. Within the last three or four years thousands of foreigners, mostly Italians and Slavs, have settled in Jefferson County. They have proved themselves highly desirable, being as they are industrious, temperate and law-abiding."[37]

The necessity of attracting European immigrants and skilled laborers from elsewhere in the United States pushed the area's largest employers to the forefront of a number of reform campaigns and arrayed them against the more conservative, fundamentalist forces in their midst. While emphasizing that "foreigners" had been "well treated in this district," the *Age-Herald* warned in August 1907 that it would "stand the corporations in good stead to see" that new arrivals were "protected against sharks and others who bear down on them." Ten days earlier it had complained that "too many worthy, sober foreigners are [being] arrested and mulcted on trivial charges" and cautioned that this practice, uncorrected, would undermine prosperity. Immigrant laborers, "being badly needed in mining and in iron and steel corporations, should be encouraged, not driven away." In October editors sharply condemned a series of police raids on an immigrant workers' quarter in Ensley.[38]

ACOA officials, who had displayed little prudence when it came to applying the full force of the law during the 1908 strike, had begun to sense the need for reform. By 1910 they were calling for regulation of the fee system in Jefferson County and had arrived at the opinion that "one of the main reasons why the district was unable to obtain labor in sufficient quantity" was "the cheapness of human life in this section." The ACOA made plans to form a committee of "men of public spirit interested in the advancement and growth of [the] District" to "develop a sentiment against the indiscriminate shooting and arresting of people . . . for trivial offenses."[39]

Birmingham industrialists assumed the stewardship of progress only so long as it did not interfere with exploitation. Industrial opposition to the fee system underscores the primacy of economic motives in employer-sponsored reform during this period and its compatibility with—even reinforcement of—the racial status quo. For many years Jefferson County's fee system had provided an effective means for checking black mobility and a lucrative source of income for local law enforcement officials. A Jefferson County grand jury condemned the fee system in 1911 as a means by which "every offense which it is possible for human ingenuity to devise for separating dollars from the poor is perpetrated." From the industrial suburb of Anniston a group calling itself the "United Organized Colored Brothers" petitioned Governor O'Neal several years later to complain that "our people and poor white peoples are not treated as . . . human[s]." They charged that "the police force do not wait for a crime to be committed" but "go and imprison a man before he commits a crime and now the station house is full." Existing arrangements rewarded authorities for such behavior; they received no salary but were paid a commission based on the number of arrests they made. The system had proven itself fertile ground for corruption and provoked the enmity of middle-class reformers, not to mention the labor movement, from before the turn of the century.[40]

In the tight labor market after 1908, however, the antifee campaign won the energetic support of industrial interests, who "complained that the county-fee officers, instead of helping them control their Negro workers, actually harassed, disrupted, and drove away their black labor force." Unlike the vagrancy laws, which were aimed at unemployed blacks and which employers continued to support,[41] the fee system encouraged police to concentrate their efforts on the gainfully employed. In cases where a valuable black employee was convicted, one study noted, "his employer would [often] pay his fine and court costs immediately in cash in order to get him back on the job." Matters were frequently settled out of court after workers posted a "cash bond," which seldom made it "beyond the officer's own pocket." The "indiscriminate arrests of working negroes for trivial offenses" had become so disruptive to industrial employers in the district that a number of them had resorted to "pay[ing] the constable in their beat to let the negroes play crap games. The corporation officials say this is the only way to keep their labor camps from being demoralized. They buy protection for their men."[42]

Another potential obstacle to employers' efforts to stabilize their labor supply arose in the form of a vigorous movement for prohibition in the district. Middle-class distress over Birmingham's growing reputation for lawlessness and murder invigorated local temperance activism, stiffening the resolve of municipal reformers, religious leaders, and fundamentalist forces to board up the city's notorious saloons and root out inebriation, particularly among the city's large black population and its unrefined industrial proletariat. Organized labor opposed prohibition as an unwarranted intrusion on workers' leisure time, but leadership in the effort to turn back the temperance forces was provided by industrial employers, not least the coal companies. Henry F. DeBardeleben, a prominent antiunion coal operator dubbed by admirers the "Christopher Columbus" of Birmingham industry, assumed leadership of the Property Owners and Good Government Committee, which led the fight against prohibition. His supporting staff included the executives of most major iron and coal companies in the district.

While anxious to condemn, alongside the antialcohol forces, the "evils attendant upon an intemperate use of liquor" and the depravities related to the "negro phase of the liquor question" in particular, the industrial opposition attempted to rally pro-drink forces with the harrowing specter of economic ruin. Prohibition would render it impossible to attract immigrant labor to the district, they reasoned, while providing native workers with yet another reason to head north. "Skilled, as well as ordinary labor is scarce and in demand all over the country," employers warned in an advertisement published in the city's main newspapers, "and we know that if a sweeping prohibition law should be enforced . . . large numbers of our best workmen would leave. There is no material at hand to replace them, and new labor, such as we need, will not come to a prohibition district." Editors at the *Age-Herald* deferred to the wisdom and experience of a high official in the Federal Immigration Bureau for a final word on the matter: "The German, Austrian, Swede, Finn, and nearly all other classes of Europeans will not go where they cannot get beer." They warned that a victory for prohibition would "cripple our industries, large and small," a message that swayed voters within city limits but was ultimately unable to stem the prohibition fever in Jefferson County's outlying areas.[43]

Within four years of the 1908 strike it had become clear that, in spite of an energetic campaign on the part of Birmingham's industrial leadership, their efforts to bring immigrants to the district had produced meager results. It was evident as well that while the immigration campaign may have been weakened by the victory of prohibition, its failure was due to more fundamental problems that the district shared with the South generally. By New South standards, Birmingham was a success story. In 1910 the city could boast of being the region's most cosmopolitan urban center. Yet heavy southern white migration to the district in the early years of the twentieth century led to an actual decline in its "foreign" white population—from 17.5 percent in 1910 to 15.1 percent by 1920. By 1912 John Fitch reported that while "immigration . . . of skilled workmen [was] greatly desired" and even "actively solicited," the European presence was "very small," and few northern workmen "could be depended on to stay after having come."[44]

The harshness and brutality imparted on everyday life by the region's caste system and the unwillingness or inability of southern industrial employers to match the wages on offer in the North were the most serious impediments to new immigration. Immigrants arriving in Birmingham hopeful that they had reached a land of freedom and opportunity were likely to have such illusions challenged quickly. The editor of the city's Italian newspaper, *L'Avanguardia,* recalled standing "speechless and stupefied, almost incredulous" at his first sight of a chain gang on city streets. The experience was enough, he recalled, to make any visitor "attracted by the reputation of [the district's] wonderful development and modernity" reconsider their decision to call Birmingham home. A Greek immigrant lured South to work in the TCI mines penned a frantic letter expressing his shock after witnessing "company police beat up a black woman because she complained that they are stealing daily wages from her husband. Then they attacked the workers with revolvers." His own situation amounted to peonage: "for the next twelve days [the company] will set us to work at gunpoint to work for nine and a half hours a day instead of eight." And on payday, he wrote, "about five hundred workers" were "shouting that the company is stealing their wages."[45]

The decision to sink roots in Birmingham was further complicated by the fresh confusion that immigration introduced among white southerners in readjusting the color line. Like their counterparts elsewhere in the South, Birminghamians found racial classification an increasingly complicated exercise as a diverse influx of Europeans began to introduce intermediary shades along the black-white spectrum. As in the Mississippi Delta, where "white schools" barred the children of Italian immigrants from attending classes, Birmingham district coal camps typically segregated Europeans of all nationalities from blacks and native whites.[46]

Even among those Alabamians who resigned themselves to the indispensability of immigration, debate raged fiercely over which nationalities would make proper southerners. When, in January 1907, Governor Comer warned against "amalgamating" a "horde" of southern Europeans with the state's largely black labor force and demanded restrictive legislation that would allow only "fit associates for our own yellow-haired, blue-eyed people," editors at the *Age-Herald* intervened with the feeble rebuke that his remarks were "a reflection on the brunettes." Most citizens of the state were not "blessed with yellow hair and blue eyes," they objected, "and the chief executive . . . should not speak so disparagingly of the majority." Nine months later editors felt compelled to enter the fray yet again to reject a call to bar southern Europeans. "Are not the Spaniards, the Italians, the Greeks and the Sicilians members of the Caucasian race and sharers in Caucasian civilization?" they asked, betraying their anxiety that for many readers the question had not yet been resolved.[47]

For immigration advocates, the delicate task of relaxing southern white vigilance sufficiently to accommodate a diverse workforce without provoking fierce reaction from defenders of the racial status quo proved too great a challenge. Immigration failed in part because the structures of white supremacy, which employers themselves played a large part in maintaining, were too inflexible to accommodate even this modest adjustment to racial custom. Few regions of the South stood to lose as much in this failing as Birmingham. Whether or not immigrant laborers were subject to the direct effects of the caste system, the standardization of brutality that its survival demanded made the Birmingham district an unlikely resting place for huddled immigrant masses seeking relief from oppression.

The fee system, fast becoming "a trap for the enslavement of white workmen as well as black," seems to have been the most objectionable of a number of unwelcome features of working-class life to newcomers. The zealousness with which religious and law enforcement authorities attempted to regulate leisure was enough to convince workers to move along. "Foreigners . . . do not like to have their amusements interfered with by a law forbidding all sorts of games in public on Sunday," Fitch wrote, while "men from Pennsylvania and Ohio can't understand why they can't play cards in public on week days." The elaborate system of barter and payment in company scrip, which dragged many Alabama miners into a cycle of perpetual debt, also struck newcomers as unjust, and a number of immigrant miners—including some employed by the district's leading coal companies—complained of being held in peonage. Restrictions on workers' democratic rights (often, as with poll tax and property qualifications, direct offshoots of the caste system) were particularly grating to those whose immigration had been spurred by the prospect of starting anew in a land where the stratification pervading European society had not yet taken hold.[48]

Although the expansive reach of racial oppression played an important role in undermining the district's appeal to immigrant workers, other fundamental problems contributed to its marked ineffectiveness. Birmingham employers were simply unprepared to match wage scales available in the North and offered few material incentives that might have dissuaded immigrants or skilled white workers from moving on. Like their counterparts elsewhere in the South, area employers "found that they could not compete in the Atlantic market. Nor could they initiate their own transoceanic labor flows because the footloose 'economic man' quality of the immigrants assured that they would not stay long in the low-wage South once they arrived." The fatal contradiction at the heart of southern immigration efforts was the employers' "cry . . . on the one hand, for only the highest type of immigrant, and on the other, to secure him at the scale of wages paid the Negro." Editors at the *Labor Advocate* expressed little sympathy for the employers' predicament. "What inducement is being offered to bring the working men to Birmingham?" they asked. "Every pamphlet which is published by our various boosting clubs holds out to the capitalist the argument that labor can be secured at a lower rate in the Birmingham district than elsewhere."[49]

With other options available to them, foreign-born laborers were unlikely to settle in the district voluntarily. The advantages they had long enjoyed in the relatively isolated regional market left local employers unaccustomed to having to compete for free labor, and their determination to preserve the region's competitive advantage by maintaining a tight hold on wages meant that fraud and coercion infused the immigration campaign. The racial order and the low-wage regime that it upheld set the tone for labor relations generally in the coalfields.

Their failure to secure a new source of cheap, dependable labor dealt a heavy blow to the operators' hopes for resolving the problem of labor supply. But that project was flawed from the beginning by their own racial assumptions, by their inability to comprehend the shortcomings in workforce performance as anything other than the reflection of innate racial deficiencies. Their identification of blacks as peculiarly unreliable is more useful as a gauge of operator prejudices than it was as a remedy for addressing the problem of workforce dependability; in some ways it represented an obstacle to transforming the situation. The logic of the operators' position led them to invest heavily in the notion that the deployment of different "racial types" would improve productivity. But for all their hopes that immigrant labor would prove the salvation of the district, by 1913 employers were beginning to admit that absenteeism rates were not substantially lower among immigrants than they were for blacks. Nor, for that matter, did native whites live up to the burden of their racial supremacy.[50]

The weak embrace of industrial discipline among workers from all ethnic and racial backgrounds points to the irrationality of the operators' racial calculations and suggests that something more tangible underlay the failure of a proper work ethic to take hold. One obvious qualification to the employers' complaints is their own inability to reciprocate the regularity demanded of their employees with regular employment. Coal operators in particular were in no position to offer steady work, and few miners could expect to work year-round or depend solely upon coal earnings to see them through. It was quite common for miners to work a two-, three-, or four-day week and to be laid off entirely for weeks or months at a time.[51] "They told you you could work tomorrow, or you could work Thursday," a black former

Pratt miner recalled, "and they worked when the company wanted to work . . . and if you didn't work, if you wasn't sick or there wasn't nothing ailing ya', they wanted you to move out of the camp." Probably a majority of Alabama miners during this period worked part of the year as farmers: "We farmed to make ends meet, and to make a living," a Gamble Mines veteran remembered. "You couldn't make a living in the mines." Native whites often worked small landholdings in the vicinity of the mines, while blacks who had emigrated north maintained close links with their roots in the Black Belt and would "drift back to the plantations" during slack times.[52]

The "inability" of Alabama blacks and whites to adjust to full-time mine employment might be more meaningfully understood as a variation on the response of agricultural people across many different parts of the world to the wrenching process of industrialization. Eugene Genovese has argued perceptively that slaveholder folklore regarding blacks' propensity to "work only long enough to provide for elementary needs and occasional debauchery" grew out of the "chasm between representatives of [an] increasingly industrial civilization and those who were emerging from preindustrial cultures and had to be disciplined, against their every effort, to a new pattern of work and a diminished idea of the value of extended leisure." The "wandering spirit" that planters and New South industrialists attributed to blacks might be understood more accurately as a rational reaction, an attempt to limit employer control over the conditions they lived and worked under. The problem from the employers' perspective was that too many blacks—and whites—shared the Italian disposition to "look around and see if there is anything better."[53]

Sustained attempts on the part of southern planters and industrialists to restrict mobility through a host of legislative, economic, and extralegal means can best be understood as a calculated effort to limit the alternatives available to laborers in a society now formally committed to, if still unconvinced of the merits of, a "free labor" system. Birmingham district coal operators were by no means strangers to such maneuvering. The extraordinary measures adopted by the ACOA after the 1908 strike to restrict miners' mobility aimed not only at restraining wage competition but also at preventing miners from reneging on debts owed their employers—including, for recent arrivals, the costs incurred by operators transporting them out of the Black Belt. Despite having imported five hundred blacks to the coalfields in 1909 and 1910, for example, the Central Labor Bureau set up by the ACOA was deemed a disaster after it was found "impossible to collect but a very small percentage of the expenses incurred," since "the parties deliberately avoided payment by moving from camp to camp." Company guards, long a stationary feature of mining camp life, were necessary not only for preventing infiltration by union organizers but also for keeping the camps free of labor agents who might lure away help with a promise of higher wages. The shooting of three such "intruders" at Corona was defended by company officials as an appropriate response to attempts "to take some of our men away from us and to other camps." Unapologetic, they vowed not to allow any disruption of their "quiet, industrious, orderly mining camp" and warned that agents "had better stay away from our place."[54]

Operators adopted measures aimed at curtailing movement within and out of the district even while conditions in their camps encouraged labor transience. In 1914 the trade magazine *Coal Age* printed an anonymous article on "Handling the Negro Miner in the South" that, while endorsing the stereotype of black labor as "naturally shiftless," admitted that restlessness was rooted in conditions for which the operators were themselves responsible. Calling for more frequent paydays (most southern mines paid on a monthly basis during this period) and reform of the commissary system, the author held that the "greatest indirect contribution" to the black miner's restlessness and "one that keeps him jumping from place to place" was the "ease with which he can be induced to borrow money . . . far beyond his means and at exorbitant rates and prices." Even among those expressing bewilderment at black mobility, therefore, a minority acknowledged that the unwillingness of blacks to "stay put" made perfect sense under the circumstances.[55]

From the late nineteenth century onward, Birmingham's leading industrial employers groped toward a two-tiered approach to the problem of labor supply. The competitive restraints imposed on the district ruled out a decisive break with their dependence on the limitless supply of cheap native labor. Native whites would continue to comprise a large portion of the mine workforce. And despite their reservations about black shiftlessness, employers acknowledged that black labor was "absolutely indispensable" to their operations. "We could not do without the negroes," one mine superintendent admitted. Even prior to the 1908 strike, TCI had undertaken a number of modest initiatives aimed at raising the productivity of its black workers.

Until 1910 or 1912 many employers had held out the hope that white northerners and immigrants could be brought to the district in numbers sufficient to fill the ranks of skilled laborers, but few projected a future in which blacks did not form an important part of Birmingham's industrial workforce. The immigration campaign never delivered the relief its sponsors had hoped for, however. Extraordinary measures could, to a limited extent, draw small numbers of immigrants into the district and stem the exodus of skilled native labor northward, but the fundamental problem—a shortage of skilled laborers willing to work "at the wages paid the negro"—could not be easily overcome. As their hopes for an imported solution faded, the leading iron and steel companies in the district, with TCI at their head, began to turn their eyes homeward, resigning themselves to the long-term development of a skilled, industrial labor force out of the imperfect human material at hand.[56]

TCI's role in introducing a comprehensive system of welfare capitalism to the Birmingham district has frequently been presented as a story of corporate philanthropy and even racial benevolence. But the historical record suggests otherwise: the new course was forced upon district employers, undertaken on a piecemeal basis as a necessary step in their campaign to recruit skilled workers. "The main thing we are driving for right now in the South," the TCI president John A. Topping had remarked in 1906, before the company's takeover by U.S. Steel, "is to get good men, skilled labor. To do that . . . we have first to present such advantages as will attract the higher order of workingmen. That means we have to regard the esthetic side and the human side a trifle more. It's good business to be human . . . and living conditions at our mines and plants must be improved, and new tenements and schools built." The company's continuing difficulties with labor supply over the next several years, compounded by the shortfall in immigration, reinforced the logic of this argument.[57]

By 1913 U.S. Steel officials sounded the alarm that "labor turnover at TCI had reached four hundred per cent per year, and the average working time per employee was twelve days of each month." How much this had to do with the unwillingness of workers to adjust themselves to full-time work is difficult to assess, but even company accounts attributed much of the absenteeism to "recurring epidemics" resulting from unsanitary conditions and lack of a safe water supply. The resulting health problems "decimated labor forces and threatened to cancel out all the advantages accruing from . . . ready access to rich mineral deposits." Those few skilled workers who had relocated to the district were generally "reluctant to bring their families into the area," and after experiencing "the unattractive and unhealthful conditions of surrounding communities," company officials lamented, they would too frequently "leave after only a short rest period." The legacy of years of neglect was beginning to exact a heavy toll on productivity.[58]

Beyond humanitarian concerns, therefore, the successful exploitation of the district's material advantages demanded a comprehensive overhaul of labor relations and the establishment of a new order in the area's coal mines, blast furnaces, and steel mills. Employer hegemony in the workplace would provide the cornerstone for such a regime. The traditional hold of craft unionism in Birmingham's steel industry, which reinforced the racial divisions between skilled white and unskilled black workers, meant that employers faced only sporadic opposition to their plans in the blast furnaces or in manufacturing. The strength of the biracial UMW meant that their early efforts would face a more formidable challenge in the coalfields, but the operators' rout of coal unionism in 1908 had removed the only viable threat to their control in the mines.

Even with the UMW out of the way, coal operators faced a daunting task in crafting a solution that would satisfy their contending labor requirements. By 1911 their realization that solid gains in productivity could not be realized without substantial long-term investment in improved technology, remedial training for native labor, and the upgrading of housing and sanitary facilities intensified pressures to reduce labor costs. The elusiveness of a quick-fix remedy to deficiencies in skill and dependability left the area's major employers with a labor problem "difficult beyond belief." In a district where, according to one account, "the original native labor body was indolent, un-trained, unreliable, and where there was no initial desire for regular employment," industrial employers were left with little choice but to "take what you had on hand and work with that," to develop a comprehensive approach to forging a sufficiently skilled, educated, dependable workforce as rapidly as possible. Tight employer control; low wages; the development of a skilled, stable body of workers out of the ragged, illiterate, and disorganized droves of black and white Alabamians fleeing the squalor and monotony of rural life: these requirements rather than philanthropic considerations would define the new order that took root in the decades ahead.[59]

The defining feature of this new order would be its continuing dependency on black labor. Not, as its prominent architects would later suggest, out of a sense of *noblesse oblige*. Nor even less out of a recognition that upon the "better class" of whites fell the burden of defending the "weaker" race from the outrages of the white rabble. Stripped of the grand rhetoric that accompanied their reorientation to native labor, the new turn was part of a strategy calculated to raise profits, and if no other group would work at "the negro's wages," then the operators would be forced to employ blacks themselves. The coal bosses were astutely aware as well that the large body of destitute and relatively defenseless black laborers at their disposal could serve as a powerful lever with which to contain the grievances of all workers, white and black. If, as the novelist Carl Carmer wrote, Birmingham was a "muddle of contradictions," a "new city in an old land," then the most glaring of its many incongruities was the chasm between the ubiquitous rhetoric of social and industrial progress and the reality of intensive, historically rooted racial and class exploitation. The new regime in the coalfields would not uproot that reality but embed it at the core of its very foundations.[60]

2. The Limits of Reform from Above

THE OPERATORS' CLAMOR for an enduring solution to their labor problems became more frantic in the difficult months and years that followed the end of the 1908 strike. The substance of their concern was not new, however. Individual mine owners had attempted to address the perennial problems of labor supply, control, and retention, with mixed results, from as early as the 1880s. The poststrike crisis merely posed familiar problems in a concentrated form and with new urgency. The desertion of the coalfields by many of the area's most experienced miners compelled operators to reckon with a set of difficulties that seemed by this time to have become permanent features of Birmingham's industrial landscape. Although the creative impulse shaping the operators' solution originated outside the district in boardrooms and stockholders' meetings far from Alabama, the cumulative experience of local operators over more than thirty years furnished architects of the new order with essential guideposts for reconstructing labor relations throughout the district.

The operators' groping attempts to develop an effective approach to managing racial composition in the mines, while occasionally degenerating into farce, were driven by a basically sound premise: their dim realization that deeply rooted racial divisions might provide a powerful lever for sustaining a low-wage regime in the coalfields. Though embellished with the trappings of progress, the new order did not abandon racial manipulation; instead it elevated the importance of race, embedding it at the heart of revamped labor relations. Traditional restrictions on labor mobility, with a lineage going back to slavery, obtained a new lease on life under welfare capitalism. Not only the more repugnant features of life in the coalfields but even many of those reforms that merited any public association of the new order with social progress had their precedents in the earlier development of the district. Large-scale construction of new camp housing and recreational facilities, along with ambitious plans for extending medical care to miners and their families after 1911, represented a concentrated, more heavily capitalized version of the initiatives undertaken by TCI's Don Bacon just after the turn of the century. And smaller operators who understood the positive effect that such investment would have upon labor turnover were limited not so much by antiquated notions of labor relations as by a lack of capital. Rhetorical embellishment concealed the degree to which the foundations of the old order undergirded the new.[1]

Three interrelated developments converged after 1908 to make it possible for mine owners to embark upon an extensive, coordinated renovation of industrial relations in the district. The principal condition for such an overhaul had been met with the defeat of the UMW. Prior to the strike, operators had complained that their "authority . . . over [the] coal mines ha[d] been growing less and less" and that their ability to stipulate "the manner in which work should be done, . . . what should constitute a fair day's work, and . . . who should be employed, had to be restored and maintained." The near-total elimination of UMW influence in the Birmingham district after the strike solved, for the time being, the crucial problem of restoring management's authority.[2]

Their ascendancy after 1908 was no guarantee that the operators could develop a common approach to the problem of labor supply, however. Beyond their shared hostility to the UMW, district operators had seldom displayed any natural inclination toward united action. They were, after all, frequently locked into competition with one another and more likely to view each other as rivals than collaborators. The experience of the strike was crucial to unifying their ranks. While involved in a bitter fight to deny their employees the right of organization, the operators came to realize the soundness of "the axiom that 'in unity there is strength.'" Although the fault lines in their ranks would never completely dissolve, most employers were persuaded of the benefits of maintaining a "compact organization" committed to "mutual aid and absolute self-defense," and the solidarity that they forged over the coming period was impressive. As the ACOA shifted from the war footing adopted at its founding in 1908 to a more relaxed posture in the years that followed, its members discovered a "broader sphere of usefulness" for their endeavors. By 1912 the Pratt Consolidated Coal Company executive George McCormack, in his capacity as ACOA president, stressed the importance of company-sponsored welfare to the district's advance and declared that "safety and harmony" had become the "principal functions" of the operators' organization. The structure and cohesion accrued during the strains of large-scale industrial confrontation would play an essential role in disseminating the gospel of welfare capitalism in the years ahead.[3]

The final condition for large-scale renovation of conditions in the coalfields, access to sufficient capital, was provided by United States Steel's purchase of TCI in 1907. One historian has observed, significantly, that the "years around 1910 form a bridge from old Birmingham into the new." To the extent that this is accurate, it owes much to the vast material resources, expertise, and industrial clout that U.S. Steel brought to bear on conditions in the district. Increasingly regarded by an awestruck American public as the embodiment of the new power of industrial capital and by friend and foe as labor's most formidable adversary, the steel trust's entry into the Birmingham district held out to area employers not only the prospect of new investment but also the cumulative experience and bottomless war-chest of the foremost defender of the open shop.[4]

Even prior to its purchase by U.S. Steel, TCI had staked out its position as the dominant force in Alabama industry. "As TCI went," one observer noted, "so went the rest of the district." In contrast to the situation in the central and northern coalfields, where wages and conditions varied widely between regions and even between individual operators, TCI "set prices and working conditions" in Alabama, "and the smaller operators followed suit." The company's domination extended to labor relations, where smaller commercial operators "gladly followed" the "strong lead" provided by TCI in their efforts to undermine the UMW. That stature was further enhanced by the arrival of U.S. Steel, which had pioneered northern-based attempts to counter trade unionism with company-directed welfare programs that preserved management prerogatives in the workplace.[5]

In trumpeting the virtues of welfare capitalism before the public, the system's leading architects invariably emphasized its positive impact on the working and social lives of industrial employees. At other times, however, they were surprisingly explicit in acknowledging that the humanitarian impulse was less significant than the system's utility in advancing antiunionism. The deep conservatism sustaining their logic was nowhere articulated more forcefully than by the president of U.S. Steel, Judge Elbert H. Gary. "'Unless capitalists, corporations, rich men, powerful men, themselves take the leading part in trying to improve the conditions of humanity,'" he warned an audience in 1911, "'great changes will come, and they will come mighty quickly, and the mob will bring them.'" To protect short-term corporate interests and guard against more fundamental challenges to the social order, Gary cautioned, responsible employers must make certain that "'the men in our employ are treated as well [as], if not a little better than, those who are working for people who deal and contract with unions.'"[6]

The implementation of welfare capitalism in Birmingham presented TCI management with a formidable challenge. Although U.S. Steel's cumulative experience provided local management with general guideposts, conditions in the district differed so dramatically from those extant elsewhere that an easy transfer of models developed in the North was out of the question. The distinct racial composition of native Alabama labor and the racial antipathy that permeated southern society offered management potentially powerful leverage against trade unionism. But an atomized workforce was one of the few assets enjoyed by TCI managers. For the most part they found themselves forced to start from scratch in bringing Birmingham operations up to par. The more extravagant ambitions accompanying welfare capitalism were at all times tempered by the urgency of upgrading delapidated physical infrastructure in the company's furnaces, mines, and mills. One hard-pressed TCI manager characterized the task facing company officials as a problem of "tearing down the old house while still living in it."[7]

In testimony before the United States Senate in 1912, the former TCI manager Don Bacon recounted the difficulties he had faced while directing Birmingham operations a decade earlier. The general backwardness of local industry can be gauged by Bacon's assertion that while the technology available to TCI in Alabama at the time was antiquated by northern standards, he considered the Tennessee company the most advanced steel and iron producer in the South. Estimating that the company probably owned rights to "a billion tons of coal," its ability to exploit these holdings was severely hampered at every turn by a lack of modern facilities. The "real value" of TCI's holdings, Bacon insisted, "was what was under ground, because there were very few things on the surface that could do their work as cheaply as was being done at [northern] properties." Bacon's successor at TCI, the New York financier John A. Topping, recalled similar difficulties. He found the property to be "run down at the heel somewhat," with "obsolete types of engines and boilers" and "generally poor equipment," and he followed through on the "rehabilitation" initiated during Bacon's tenure before selling the property to U.S. Steel.[8]

Although the labor problem figured prominently in forcing TCI's hand, it constituted only one of a range of factors shaping the modernization program embarked upon by company officials. Many of the specific improvements pursued under the rubric of company "welfare" (such as the introduction of mining machines and the construction of new blast furnaces) might more properly be classified as necessary upgrading of an archaic industrial infrastructure. Moreover, the creation of a distinct category of expenses related to welfare seems to have allowed employers to imbue what might otherwise be regarded as elementary obligations with an air of benevolence and philanthropy. "The number of items which U.S. Steel lists under 'welfare,'" one critic pointed out some years later, was "really imposing": included among such listings were "accident prevention, relief for injured men and families of men killed, sanitation. . . . The installation of urinals and . . . lockers . . . ; the investment which the company has made in [housing] is likewise so classified. Piped systems for drinking water are part of welfare, as are the salaries of the full-time safety officials."[9]

Given their difficulties with labor supply and efficiency and their growing realization that the district's long-term prosperity required substantial investment, the commitment of a number of key operators to spreading TCI's model of company-sponsored welfare throughout the district is unsurprising. The business logic upon which it was based seemed compelling and sound. The potential relief it offered from chronic crises, like the promise of immigration only a few years earlier, seemed too tempting and economical to pass up. By 1912 or 1913 the ACOA had been converted to the redemptive potential of reform from above, and a number of prominent operators proselytized on behalf of the new order.

How profoundly the doctrine of welfare capitalism had permeated their outlook can be gauged from remarks made by the ACOA's secretary-treasurer in his 1913 annual report. In words that could have emanated from the lips of Judge Gary, the official boasted that among the important advances made over the previous year none seemed more significant than the operators' success in making it a matter of "policy" to "get their [sic] first with a square deal and to remove all just causes for dissatisfaction." Impressed by the energetic work undertaken by TCI over the previous two or three years[10] and prodded by Tennessee company officials, the ACOA took its first steps toward embracing company welfare in 1911 when it engaged the services of the noted sanitation expert Morris Knowles. The organization commissioned Knowles to complete a "special study [of] sociological conditions surrounding the mine workers of Alabama . . . to embrace sanitation, housing, hygiene, and amusements." The study was to include a "full detailed report covering every coal mine represented in the Association" and would be disseminated to every operator-member in the district.[11]

Two key developments over the following year combined to accelerate TCI's efforts and reinforce the ACOA's evolving commitment to company welfare. The first was TCI's forced abandonment of the state convict lease system at the end of 1911. In her charitable biography of Judge Elbert Gary, Ida Tarbell attributes the change to the steel executive's deeply held abolitionist convictions. "When the contract with the state for convicts ran out," she recounts, George Gordon Crawford took up its renewal with Judge Gary. "'Think of that!'" she recalls hearing Gary protest. "'Me, an abolitionist from childhood, at the head of a concern working negroes in a chain gang, with a state representative punishing them at a whipping post!' 'Tear up that contract,' he ordered, 'It is not necessary to consult anybody. I won't stand for it.'"[12]

The truth seems considerably more complicated. Whatever Gary's moral objections, TCI had for many years been the state penal system's most substantial customer, and Crawford himself had boasted that convicts mined the "cheapest coal in the state." In common with large-scale operators throughout the district, TCI management was mindful of the advantages that convict labor offered them. Along with the guarantee of an uninterrupted labor supply, convict labor handed operators a powerful means of combating trade

Convict burial place outside Pratt Company's Slope No. 12. This photograph, featured in John Fitch's January 6, 1912, exposé in *The Survey*, was accompanied by the explanation that "with neither scripture nor prayer, deceased prisoners are placed in the soon obliterated graves in the woods." (Courtesy of the Robert D. Farber University Archives, Brandeis University, Waltham, Mass.)

Occasionally, free and convict miners (in the white uniforms) worked side by side, but more often operators deployed them in different sections of the mines. Free miners complained that they were assigned more "dead work" while convicts were put to work where coal could be extracted with less labor. (catalog no. 24.32; courtesy of the Birmingham Public Library Archives, Birmingham, Ala.)

union organization and enforcing low wages among free miners. James Bowron, in earlier years a leading TCI official, dismissed the "constant charge, bandied about to and fro in the press, of inordinate profit received from the labor of helpless men to satisfy corporate greed. There is no profit in convict labor." The only advantage derived from the use of convicts in the mines, according to Bowron, was "continuity of work free from strikes." Besides the guarantee of continued output, the very presence of convicts at work during stoppages imparted confidence to those free miners inclined to scab. "When trains of coal are seen moving," Bowron reasoned, "this is an encouragement to other men who realize that the strike is not a complete one and that they may participate in the mines without undue feeling of persecution."[13]

Even during periods marked by industrial peace, Alabama's geological conditions presented mine operators with an added incentive for pitting convicts against free miners. Unlike the northern coal seams, whose thickness typically varied between nine and fourteen feet, Birmingham district seams averaged between thirty-six and forty-eight inches. This made it necessary for the Alabama miner to work "from sunrise to sunset to load about seven tons a day, while the miner in Illinois could load from eight to ten tons" at a higher rate. Compounding the disadvantage was the natural inclination of convict operators to "[work] convicts in the most productive sections of the mines, whereas free miners, who worked on a tonnage basis, were assigned to sections which required considerable 'dead work' before they could begin producing coal." TCI's resistance to the removal of their convict labor was therefore unsurprising. Prior to termination of its contract with the state, the company's two convict mines were "consistently the most productive" of its vast holdings.[14]

The termination of TCI's contract with the state therefore had very little to do with management's abolitionism and evoked a "very vigorous protest" from George Crawford. Although in hindsight he claimed a commitment to "cease working convicts at such time as free labor could be installed in their place without detriment to our operations," Crawford admitted that the company's "mines and mining operations were to quite an extent dependent upon them" and apparently had every intention of renewing the existing contract at the end of 1911. Through bureaucratic oversight, TCI officials failed to submit the required paperwork in time to meet the annual deadline. State officials, prodded by TCI's competitors at Pratt Consolidated and possibly bearing a grudge against TCI over the "persistently low convict lease rates after 1908," refused Crawford's frantic appeals and moved the more than three hundred convicts employed at the company's number 12 mine to the Banner mine operated by Pratt.[15]

Although events leading up to the removal of state convicts from TCI mines are disputed, there can be little disagreement over their overall impact on the company's commitment to welfare capitalism. The unexpected withdrawal of convicts compelled Crawford to expedite plans for filling their places with competent free labor and forced TCI further along the path of modernization. Crawford complained to state officials that, with barely a month's notice, "no adequate force of free labor can be gotten together to work the mines, nor can the houses be built to take care of them." Extensive renovations would be necessary before normal production levels could be resumed. "Many houses must be built, a permanent mining town provided, and arrangements made to secure the labor." The abrupt end of convict lease therefore provided TCI officials with a powerful, if unwelcome, impetus to overhaul their mining facilities.[16]

Simultaneously with the crisis at TCI the ACOA as a whole found itself in an equally difficult predicament—one that accelerated their overall commitment to the new order while revealing tensions and fault lines within their ranks. The immediate catalyst for the crisis was a special series of articles in the national magazine *Survey* authored by John Fitch, Shelby Harrison, and others in January 1912. On the whole, Fitch's feature article on Birmingham industry seemed critical but fair: he commended employers who had undertaken modernization while chastising those that lagged behind. The overall impression conveyed in the report was not flattering to the district, however. Fitch acknowledged that the district's outlying mining camps had "much to recommend them." "Often they are built on cut-over timber land," he explained, "where many trees are still standing and where there is considerable natural beauty." On close examination, however, the newly constructed company housing was "desperately cheap" and frequently "unpainted and weather-beaten." Those operators who had turned a deaf ear to the gospel of progress fared much worse. The Sloss-Sheffield company came in for particularly harsh condemnation, its workers' quarters inside the city limits being described as "an abomination of desolation," where the "houses are unpainted, fences are tumbling down, a board is occasionally missing from the side of a house." By way of explanation, the company president, Col. John C. Maben, was quoted saying that he did not believe in "'coddling workmen.'"[17]

The reaction to Fitch's exposé from members of the ACOA varied from bitter condemnation to reluctant admission that the report contained more than a grain of truth. The president of the Empire Coal Company, Walter Moore, who in the postwar period would assume leadership of the operators' organization, wondered why the Birmingham district had been "entered in the night, as it were, and lambasted by men whose minds never light upon anything concerned with the good, the beautiful, and the true?" "There is a handful of men in Birmingham who have withstood the fever's blistering breath," he wrote in the *Birmingham Age-Herald*. "They have conquered the elements and brought wealth and happiness to many score of thousands. Yet they are crucified before the world by a staff of alleged philanthropic men whose whole life is spent turning up the muck." Colonel Maben of Sloss-Sheffield considered the article "the most inexcusable injustice that had ever been done a District," and the incident crystallized for him and a number of skeptical operators the bankruptcy of the "new-fangled ideas" animating TCI's turn to welfare capitalism. By month's end the Sloss manager J. W. McQueen had resigned from the ACOA executive committee and four coal companies, including Sloss-Sheffield (the third largest coal producer in the state), Moore's Empire Coal Company, sixth-ranked Alabama Consolidated Coal and Iron, and the Davis Creek Coal and Coke Company, operated by Maben's son, had withdrawn from the operators' organization in the first dramatic sign of a break in their ranks.[18]

For the majority who remained in the ACOA, the fallout from the *Survey* series only confirmed the need to quicken the pace of reform. Their reasoning was simple: members feared that unless vigorous measures were adopted to counter the adverse publicity arising out of the investigation, they would be unable to induce skilled labor to settle in the district. "If we are to find the men for a twenty-five to thirty million tons annual output," Ben Roden of the Roden Coal and Coke Company at Marvel warned his fellow operators, "we must have good camps and make them attract the labor we desire. A large number of employees who have been with you for a number of years is a decided asset. They are satisfied, and soon your camp means home." In testimony before Senate hearings later that year, John Fitch reported that despite the defections, the district's "leading coal companies" were sticking to the movement, highlighting the efforts being made by Woodward Coal and Iron, Republic Iron and Steel, and Pratt Consolidated to raise standards in the mining camps. The gravity with which some of the central figures in the ACOA regarded their predicament is confirmed in the overtures they had made several months earlier to Fitch himself, soliciting his assistance in appointing a qualified individual to oversee their planned sanitation study.[19]

TCI's struggle with state authorities over access to convict labor along with the fractures in the operators' ranks exposed in the *Survey* feature illustrate the need for caution in making broad assumptions about the spread of progress in the coalfields. In the first place, the operators' formal embrace of welfare capitalism did not necessarily translate into dramatic changes in the day-to-day lives of district coal miners. Even with the best of intentions, the owners of smaller and medium-sized mining operations frequently lacked the material resources that made large-scale improvements feasible at properties run by the district's industrial giants. Alongside the model camps being constructed lay a great many others "built without regard to permanence or comfort," which Fitch regarded as "product[s] of a period not yet entirely past in the Birmingham district, during which operators have had a desperate struggle to get sufficient capital to develop their mines."[20]

While striving to project satisfaction and unanimity before the general public, the operators were well aware of the disparities developing between advanced operations and those unable or unwilling to implement improvements. Mindful of the importance of showcasing progress in the leading camps, however, they found a sympathetic business press and a coterie of pliant state officials willing to indulge their claims. When the state mine inspector issued an upbeat annual report in 1912, miners flooded the pages of the *Labor Advocate* with letters complaining that he had "convey[ed] a misleading impression to the public . . . that the miners' life in Alabama is a perfect paradise." "It is true," the UMW District 20 organizers J. L. Clemo and William Harrison wrote, "that the Tennessee Company at its new openings have installed modern machinery and have built some very good houses . . . but even they haven't gone to the extent that [chief inspector Charles H.] Nesbitt would have you believe." Threatening to "show the public just how much or how little of this reform work has been installed in this district," union officials called for an investigation by a "disinterested board of inquiry," complaining that Nesbitt was "using his good offices . . . to perpetuate and boost the gross misrepresentations of the operators' association." The report had obviously been "gotten up," another article concluded, "for the purpose of attracting more miners into this territory to share the miserable conditions of most of the miners who work in this district."[21]

Almost every letter critical of Nesbitt conceded that a few individual companies had undertaken substantial renovations. "We want to give credit to whom credit is due," wrote Clemo and Harrison, and their own letter commended the UMW's arch rival, TCI, for the improvements underway in several of its camps. Even the *Labor Advocate* acknowledged that TCI's ore mining camps at Reeders, Fossil, and Ishkooda were "models, and have improved wonderfully the condition of the employees of the company" and noted that Ben Roden's camp at Marvel was being transformed into "an ideal camp of this character."[22]

The miners' main complaint, however, centered on the selectivity of the inspector's report: Nesbitt's account presented a few exceptional operations that had undertaken ambitious reform as representative of trends in the district as a whole. "The commissary system is extolled because some of the companies run their commissaries in an exemplary manner. Because some few companies have decent housing conditions Mr. Nesbitt makes them . . . all have the same conditions." One miner penned a sarcastic report on his visit to the Warrior–Black Creek Coal Company camp, owned by the ACOA president G. B. McCormack. "This organization stands for the uplift and better housing of the coal miners," he wrote. But "the houses they live in are scarcely fit for cows and mules. The partitions dividing the rooms are sealed only on one side. There is no ceiling overhead—only the roof of the house. All you have to do to see what is going on in the other room is to grab a two-by-four, pull yourself up and look over. This is what they call uplift." Another protested that "there are hundreds, if not thousands, of men, women and children in Alabama's mining camps in need of food and clothing, yet the

[ACOA's] news and photo agents . . . picture nothing but plenty and contentment." And a casual report from the coalfields several years later, in which a miner "thank[ed] God for his fine blackberry prospects," suggests that the deprivation detailed in such reports was authentic. "You don't realize what a blessing this is to poor working people," this correspondent explained to readers of the *Labor Advocate.* "These people live on blackberries and bread for days at a time."[23]

Standing to benefit from the favorable impression left by Nesbitt's report and the upbeat reviews its efforts had received in the business press, the ACOA chose to remain above the fray, taking no part in the public controversy. In internal discussions, however, operators acknowledged that reform work was proceeding unevenly and that tangible improvements were more elusive than good press. The most striking repudiation of the state's claims could be found in those substantial operations that resisted the new management style in principle as an affront to custom. The most prominent and energetic defender of the old order was Sloss-Sheffield, which, under the cantankerous leadership of Colonel Maben, "remained as committed as ever to its traditional southern ways." "While developments at TCI became increasingly congruent with progressivism," W. David Lewis has observed, Sloss's "aging, ultraconservative leadership marched out of step with it. Maben and other corporate officials not only refrained from embracing welfare capitalism but actively opposed it."[24]

The main fault line among district operators divided advocates of employer-directed reform, led by TCI, from a "small group of employers" characterized by Fitch as being "without ideals and apparently without shame . . . exploiters both of natural and of human resources [who] run their mines with a minimum of capital and a maximum of danger, intent only on output and speedy profits." One such operator boasted that he "had found a way to prevent labor troubles" by "hiring no one who knew how to read." When asked by Fitch whether he had adopted this method in all of his mines, the operator complained "regretfully" that he "couldn't." "There aren't enough illiterate niggers to go around," he fretted. "They're spoiling them now-a-days by educating them."[25]

The attention devoted by chroniclers of this period in Birmingham's development to the "progressive" companies has left the impression that oldstyle management methods were rendered anachronistic, but they were not an aberration at all. Not until the 1920s, after substantial injections of new capital, would Sloss-Sheffield undertake "a modernization program in any way comparable to the one that took place at TCI after 1907." In the meantime its executives "pursued an increasingly conservative course by conducting some routine repairs and . . . installing only a modest amount of new equipment." Moreover, the rearguard action at Sloss seems remarkable only because of the company's prominence and the deliberation with which it was carried out: a host of smaller operators held stubbornly to time-honored methods.[26]

Many smaller firms were simply unable to finance substantial reform. Milton Fies, the superintendent at Henry T. DeBardeleben's Sipsey mine in Walker County, claimed some years later that his camp, "when first completed in 1913, was one of the show places of the state as a mine village." Plans for building a new school for white miners' children at Sipsey were nearly suspended, however, when funds ran out before construction was finished. At Fies's urging, DeBardeleben issued funds to finish the building, but black children at Sipsey attended school in the "ramshackled, one room . . . humiliating . . . structure" of the old school for several more years before improvements came their way. Plans for new housing proceeded slowly at Sipsey, and in 1920 Fies still considered himself "greatly handicapped" in raising "output" by the shortage of proper housing. Fies's comments suggest that half-measures sometimes produced more resentment than no effort at all. When the Alabama Fuel and Iron Company installed a "picture show and auditorium" at Margaret, the improvements "caused some jealousy and rivalry . . . at Acmar," forcing management to "do the same thing for the Acmar people." Temporarily strained by the costs of having to erect a new building, company officials nevertheless predicted that the "investment" would pay off by "keeping labor satisfied."[27]

The unevenness visible at the local level characterized the district as a whole, contradicting the image put forward in ACOA literature. As late as 1922, some fifteen years after TCI had embarked upon modernization, the U.S. Coal Commission found that of 713 company-controlled coal mining communities nationwide, only one in Alabama ranked among the highest seventy-five. Three scored in the lowest group, and less than 4 percent provided running water. None offered bathtubs, showers, or flush toilets, and less than half were equipped with electricity or gas. Government investigators reported that open sewerage and lack of fresh water led to typhus epidemics "from time to time—every summer." At least one mining superintendent housed his workforce within sight of the company store "so [the] tenants can be watched" and "consider[ed] the fact that there [was] an independent, incorporated town adjacent a matter for regret."[28]

During the same week that readers of the *Labor Advocate* challenged Nesbitt's account of conditions in the mining camps, ACOA officials heard their secretary-treasurer admit in his annual report that the welfare work envisioned by the organization was "barely in its inception." "Several . . . members had their operations in fairly good condition before investigation," he remarked, "yet, I regret to report that there is much room for improvement everywhere." The common theme struck by ACOA officials at every level was the necessity of closing the gap between the advanced companies and those lagging behind. "It was formerly thought that any kind of shack, giving crude shelter, was good enough for the laborer," Morris Knowles had told an audience of operators a week earlier. While lauding the ACOA's efforts to retire such attitudes, he complained that "there are still a few employers who cling to such a conception of the relation of master and servant." In camps where welfare work had taken hold, officials observed, it had shown itself to have "a salutary effect on the disposition of labor, and closed that many openings for attack." But the slow pace of renovations and the resiliency of traditional methods jeopardized the industrial peace settling over the district. "We can never reasonably hope to be immune from trouble," ACOA officials warned, "until the laggard has forged abreast of the leaders and we present a solid phalanx."[29]

Welfare capitalism in district mines was therefore distinguished less by the tangible benefits it delivered to workers than by the unevenness with which it was implemented. More striking even than these limitations, however, were the sharp boundaries within which such work was confined, even in those camps run by the system's most avid promoters. The new order had been ushered in, after all, not as a philanthropic experiment but with the explicit objectives of raising productivity and increasing long-term profits. This is not to deny that the turn to welfare capitalism delivered "much-improved . . . conditions for thousands of Alabama's most disadvantaged citizens," as Lewis has suggested, or that individual operators conceived their new commitment in humanitarian terms. Many saw themselves as the local embodiment of the progressive impulse then sweeping the nation—one that was aimed in part at purging industrial relations of bitter conflict. No doubt many mine owners considered the dual objectives of improving conditions and enhancing their competitive position entirely compatible. But the particular context in which welfare capitalism took root in the district—following close on the heels of a bitter period of protracted, chronic labor trouble and driven by the necessity to hold down wages while cultivating a skilled, dependable labor force—left an indelible stamp on the foundations of the new order.[30]

Two features in particular stand out about the public face of welfare capitalism. While the ACOA did little to counter the impression that reform of the district coalfields derived from philanthropic rather than material motives, among themselves almost every attempt to win converts to the new order was framed in terms of its impact on profitability. In measuring the impact of reforms in his annual report for 1912, the ACOA's secretary-treasurer emphasized the fact that "where the work has progressed to any extent, the output and number of working days per man has been increased, and the number of accidents decreased." Dr. Lloyd Nolan, the director of TCI's health services, stressed that "sanitation is not only a moral responsibility, but a work that pays, for the prevention of disease is always cheaper than its cure. The sickly workman, or the workman with a sickly family, is an expensive investment." Marion Whidden, another TCI official recently appointed to oversee the company's social work in the coal camps, explicitly rejected the notion that "philanthropy" was behind the new interest in company-directed recreation work. "The sums expended for the promotion of physical, mental, and moral welfare of employees result in a greater profit to the employer," she explained in a speech before members of the ACOA. The attempt to disassociate company welfare from philanthropy represented less a call to unbridled profit-making than a recognition of the inherent danger of its lapsing into overt paternalism. Nevertheless its architects measured every element of company welfare, from housing and safety to recreation, hygiene, and the cultivation of domestic tranquility in the camps, with careful and deliberate calculations as to their effect on profits.[31]

The other distinguishing feature of the new order was its hostility toward any manifestation of independent initiative on the part of workers. However lavish the improvements instituted at a particular mine, they were invariably directed from above, usually without any pretense of input from miners or their wives. Even among those employers at the forefront of innovation, Fitch noted, "there is something tremendously lacking." "Cooperation between employer and employee . . . is limited to safety work and the testing of mine scales—a somewhat compulsory combination after all—and to 'suggestion boxes' into which an employee may drop a note setting forth his grievances." An astute mine boss, attuned to the sentiments of those working under him, might be adept enough to anticipate and resolve minor grievances as they arose. But any attempt by miners to press their demands through independent organization was absolutely precluded in the new arrangement. Judging from operators' reactions to the few rare breaches in the new protocol, such initiative smacked too much of insolence and revived, in their eyes, the specter of trade unionism: thus their emphasis on "getting there first with a square deal." "It is better to provide and maintain these safeguards voluntarily," explained an ACOA official in a classic rendering of the paternalist logic, "than to have them forced upon you by the law or public demand." To this he might have added: "or by industrial action on the part of employees."[32]

The operators' willingness to undertake improvements was predicated upon workers' acquiescing in the direction of welfare work from above. Given the collapse of the UMW and the desperate conditions prevailing in the coal camps during this period, this was a trade-off that large numbers of miners were willing to entertain. But acquiescence should not be mistaken for enthusiasm. However much the architects of welfare capitalism endeavored to project their arrangement as the embodiment of democratic progress, the new order was not the result of a negotiation among equals, and compulsion suffused every aspect of company welfare in the coalfields. Employer-directed reform therefore pursued a limited amelioration of conditions while reinforcing the authority of mine operators over their workforce.

Some approximation of the boundaries of progress-from-above can be gleaned from the operators' approach to mine safety during this period. For many years Alabama miners had complained about the lack of uniform safety standards in the state's coal mines. Two serious mining accidents in 1910—an explosion at Mulga that killed forty-one miners and another at Palos that took the lives of ninety-one—raised the grim tally of fatalities in that year to a high of 238, or an average of twenty deaths per month, reviving public debate over mine safety. The Palos explosion in particular and a series of scathing articles in the *Labor Advocate* charging operators with profiteering from the disasters roused miners and the general public to press for effective mine safety legislation.[33]

After the Birmingham Trades Council organized a "Tag Day" to raise funds for the widows of those killed at Palos, Mrs. Frank Harns, who directed the relief effort, reported having turned down a request by the Red Cross to "hand over" the funds raised by union sympathizers. Harns protested that the "money distributed by [the Red Cross] was not paid direct to the sufferers but orders were given them on the Drennen commissary [where] supplies obtained on these orders were procured at most exorbitant rates." She charged that "poor people were compelled to pay ten cents for five cents worth of crackers and that other necessities were sold at the same ratio of robbery." In a revelation that angered BTC supporters even further, another of the women involved, Oleana McNutt, complained that one Red Cross representative had "tried to get some of the women to give their children away, but was unsuccessful." McNutt did her best to dispel the widespread assumption that "the Miners' Union is dead in Alabama." "If some of you would go to Palos, you would change your mind," she predicted. "The mothers there are training their little sons in unionism and they say if there had been a Miners' Union, the explosion would probably never have occurred."[34]

Sensing an opportune moment to apply the dictum of "getting there first," the ACOA's executive board appointed a Committee on Mine Casualties to "undertake the framing of a more efficient mining law for the State." Obtaining the sponsorship of the state senate president Hugh Morrow, chief legal counsel for Sloss-Sheffield and "probably the most vigorous legislator" at Montgomery,[35] the ACOA managed to have its proposal for the mining regulation read—without any disclosure that they were the architects of the bill—before the legislature in early January. When the substance of the operators' proposal became public, UMW officials objected that it was designed to relieve mine owners of all liability for accidents and launched a vigorous campaign to oppose the bill. "If it becomes law, we do not see how any self-respecting miner can come to Alabama," they protested. "There are many penalties imposed by the act, but practically all of them are aimed at the miners, and only one . . . at the operator." The District 20 president J. L. Clemo denounced the bill, declaring that it would work "solely for the protection of the operators against any liability for the maiming and killing of those whom they send down into . . . the earth."[36]

UMW officials considered the bill so one-sided that they preferred to forego any new legislation rather than see it passed in the proposed form. "Kill it," urged the *UMW Journal,* "and don't disgrace the state." The bill languished untouched for several months until it was amended slightly and reintroduced by John D. Hollis, the chairman of the House Committee on Mining and Manufacturing and a store manager for the Pratt Consolidated Coal Company. The revised bill was finally passed unanimously on Friday, April 7, but almost immediately legislators troubled by certain provisions vowed to reconsider it the following week. On the following day, the worst mine accident in Alabama history occurred at Pratt's Banner mine. One hundred twenty-eight men perished in a gas explosion—all but five of them convicts and all but ten of them black. The charged atmosphere following the Banner explosion tempered the operators' determination to force through undiluted safety legislation and opened the road to compromise. In its final version, the bill remained heavily weighted in their favor, however. While miners' representatives succeeded in augmenting the number of inspectors assigned to the mines, the operators prevailed in barring the release of accident reports to the public and in parrying last-minute attempts to assign liability for accidents caused by employer negligence.[37]

The operators regarded their success in framing specific provisions of the new law to be of secondary importance to their overall ability to deflect the

Crowd at the entrance to the Pratt Company's Banner Mine, where 123 convict miners were killed by an explosion in April 1911. (reprinted from *The Survey,* Jan. 6, 1912; courtesy of the Robert D. Farber University Archives, Brandeis University, Waltham, Mass.)

threat to their authority in the area of safety. Having thwarted a major attempt to regulate the industry, leading operators initiated mine safety and rescue programs that would alleviate public concern by eliminating the most flagrant safety violations. TCI established its own cadre of mining inspectors and reportedly "adopted the most rigid safety rules." In each of their mines, John Fitch reported, "there is a rescue corps trained in first-aid and mine rescue work. A competent man secured from the . . . Bureau of Mines is employed to have charge of the rescue work and to train the first-aid corps." Operators organized interdistrict competitions between (segregated) mine rescue squads, and safety figured prominently in ACOA lectures and literature. The results, however, like the spread of company welfare in general, were uneven. The operators' success in shifting the burden for compliance onto individual miners meant that safe conditions could only be had by sacrificing wages. Miners bore the increased expense of using the safer permissive powder in place of black powder, for example. And where wages were pinned to productivity, miners would inevitably continue to cut corners. "Dead work" such as bracing or dust removal was "charged against the cost of the coal," complained one miner, "which makes the average mine foreman hesitate about doing such work." It is not surprising, therefore, that while safety seemed to improve at mines operated by the largest companies in the district, overall Alabama's fatality rate remained well above—and frequently twice as high as—the national average.[38]

Critics charged that the minor compromises that had been effected in the final bill were more than compensated for by the powerful influence that operators continued to wield in the selection and conduct of state-appointed mine inspectors. A week after the carnage at Banner, an engineer from the Birmingham Water, Light, and Power Company wrote Governor Emmett O'Neal upon learning that his application for the position of chief mine inspector had been rejected, apparently at the insistence of the ACOA. While harboring no "personal feeling against the mine Operators for their anxiety to keep me out of office," the applicant denounced the owners for their "cowardice of intelligence," suspecting that his fate had been sealed by their knowledge "that they could not control me."[39]

While Nesbitt's appointment as chief inspector appears to have produced no such anxiety for the operators, miners complained that the safety routine instituted by him was a farce and that mine management knew well ahead of time when inspectors were coming. Preparing for a mine inspection was "just kinda' like getting ready to go to church," one miner recalled. "You get your best clothes and put them on, you wore them to church and when you come back, pull them off and that's about the way it was with the company . . . and the mine inspectors." A black former Pratt employee complained that inspectors and the companies were "hand in hand 'cause the company sent a man with him to take him around. Just a blind dog in the meat house, that's all he was." A group of miners from the Douglas Coal Company near Jasper prevailed upon a representative of the Farmers' Union there to write the governor and complain that Nesbitt was "not doing his duty as the law requires." Their employer was "paying no attention to the laws passed by the last Legislature," storing explosives in an "old dilapidated dwelling house" and "cheating their men" through inaccurate weights. Miners complained that a state official held a one-third interest in the company, and their petitioner was so wary of state collusion with company officials that he pleaded with O'Neal "not [to] refer this to the Inspector's Office." "Send your agent to me first to look over the property," he urged, or "say nothing about this."[40]

Mineworkers displayed little faith in the state's commitment to reform and relied instead upon their own power in the mines to enforce existing legislation. The failure of state authorities to rectify chronic complaints about the shortweighing of coal demonstrated the necessity of workplace organization and the unwillingness of the operators to tolerate any challenge to their prerogative. Technically, miners had won the right to elect a checkweighman at each tipple to guarantee the accuracy of company scales and ensure that they were being properly compensated, and they considered the clause one of the few tangible gains they had won at Montgomery. Almost immediately after the bill's passage, however, its implementation was sharply contested by management. At many mines operators either refused to allow the appointment outright or insisted on selecting the checkweighman themselves. When, in December 1912, miners at Piper attempted to elect a checkweighman and pay him the equivalent of six tons per day (to be deducted from their own wages), their choice was rejected by the mine superintendent, who "ordered the mines to close down . . . and ordered the miners to take their tools home with them." The Piper workforce stayed out of the mines for a week and organized a delegation to meet with management the following Sunday. After securing a reversal of the decision, delegates called a mass meeting for Monday and miners returned to work the next morning.[41]

The outcome at Piper was exceptional, however. At most mines, union officials protested, "the superintendent or mine foreman . . . call or attend [*sic*] meetings and act as chairman of said meetings. Unless a man nominated by [miners] . . . is satisfactory to the company the meeting is adjourned and no election is had." Where the possibility of organizing industrial action along the lines of the Piper incident was out of the question, there was little that individual miners could do to assert their rights. "If any miner makes any kick about the method of electing checkweighmen," the UMW organizer W. R. Fairley observed, "he [is] discharged forthwith."[42]

The ACOA's response to the weights controversy was driven less by pecuniary concern over potential losses than by their determination to resist the more fundamental challenge to mine management's authority. Individual operators might deliberately rig weights to their advantage, but any medium- to large-scale operator would have been foolish to adopt such flagrant means when there were less transparent methods available for accomplishing the same ends. The very existence of elections independent of company control or the mere act of sending a committee to discuss working conditions with management seemed much more serious matters, however, and smacked of an attempt to resurrect the UMW. The dispute over checkweighmen seemed to ACOA officials to coincide too closely with a recent visit of UMW officials to the district, and they perceived—perhaps correctly—an attempt to revive the UMW's fortunes through the weights controversy.

Operators feared that the union had managed to preserve an underground network in some areas, and if this were true, the appointment of union stalwarts as checkweighmen would provide an ideal opening for reorganization of the UMW. Checkweighmen maintained daily contact with virtually the entire mine workforce. The position therefore provided an excellent vantage point from which to chronicle the grievances and gauge the sentiments of rank-and-file miners and an unparalleled network for communication throughout the mines. To guard against the UMW's revival, the ACOA voted to step up surveillance work and place it under the central supervision of their secretary.[43]

Within a few short years of the 1908 strike the basic contours of the mine owners' much-vaunted safety policy were clearly discernible. Much to the dismay of district miners, operators had successfully resisted a broad-based attempt to expand state regulation of the coal industry. With the threat of outside interference neutralized, the larger companies initiated comprehensive mine safety programs that were in certain respects among the most advanced anywhere in the country. The ACOA attempted to generalize from these few success stories and spread their example across the district. For the most part, however, laissez-faire continued to be the watchword of Alabama mine safety. This meant that safety continued to be enforced at the discretion of individual mine owners and that miners would absorb much of the cost for enforcing it. The absence of effective organization left miners "at the mercy of the earth" and meant that injury or death on the job brought destitution for the hundreds of widows, orphans, and maimed workers that the Alabama coal industry produced on an annual basis.[44]

Few incidents illustrate so poignantly the dark underside to laissez-faire "progress" as one revealed in a letter from Hugh Grant, the supervisor of Alabama's Industrial Rehabilitation Board, to Milton Fies, the superintendent of Sipsey mine. Grant wrote Fies to inform him that he had come across a former Sipsey employee named George Lawson on the streets of Selma. Grant found Lawson, blinded by a mine accident, "sitting out on the sidewalk on a very cold, rainy night" in a "very pitiable condition." Without contemplating that the company might help support its former employee, Grant inquired whether Fies would organize a collection among "some of the men who knew Lawson at Sipsey" to defray his living expenses while taking a state-sponsored course in mop making. A similar incident discloses even more strikingly the colossal disadvantage suffered by miners in pressing claims for compensation. A black miner named Will Pond had been badly burned in a mine accident at Sipsey. Fies wrote his superior, Henry T. DeBardeleben, in 1918 to advise him that the company lawyer "thinks he can settle with the negro . . . for $100." He instructed DeBardeleben to bring along "$125 *in one dollar bills* and a release" when he visited Sipsey the following week. "Please don't forget," Fies stressed. "The negro's in the notion." The unwillingness of state authorities to force a more equitable arrangement upon the operators left men like Pond with no alternative but to settle on the pathetic terms dictated by their employers.[45]

The limitations of employer-directed reform are likewise apparent in the ACOA's unwillingness to tamper with the odious convict lease system. In the first place, the transfer of more than three hundred convicts from TCI's number 12 mine to Pratt Consolidated's Banner mine did not seem to unduly affect George B. McCormack's tenure as head of the ACOA, despite that organization's ostensible commitment to "uplift." More importantly, however, TCI's abandonment of convict lease coincided with an expansion of the system on another front. Significantly, the single case in which mine owners introduced convicts to displace free labor in the post-1908 era occurred during a strike against the Montevallo Coal Mining Company in 1913. The mine, located at Aldrich, was "one of the oldest" in the state and "had always been worked by free labor" prior to 1913. It was also, coincidentally, the only remaining mine at which the UMW had managed to hold together a union local and maintain a contract after the 1908 defeat. In 1913, however, the mine was purchased by two members of the ACOA, who were determined—and very likely prodded by their peers—to operate the mine on a nonunion basis.[46]

In early September, nine months before the existing union contract was due to expire, the new owners issued an ultimatum in which they threatened to "put contractors in the mine, and [if] necessary, work it Non-union." After protracted efforts on the part of District 20 officials to negotiate an understanding with the new management the UMW finally called a strike at Montevallo to commence on January 1, 1914, and strikers "were immediately replaced by convicts." Whether this had been the intention of the new owners all along is difficult to determine, but the maneuver certainly fit with the ACOA's efforts after 1908 to rid the district of any remaining pockets of UMW influence.[47]

When, on occasion, individual operators contemplated a break with convict lease, they seemed less troubled by the system's moral dimensions than by the disadvantages it imposed on those working their mines with free labor. Henry T. DeBardeleben launched an energetic campaign to "eliminate the convict from the mines of Alabama" in the spring of 1915, but his appeal to fellow operators rested on the "effect of convict coal competition in our present coal market" rather than on the brutality of the system. The mixed response that DeBardeleben's appeal elicited from mine owners around the district illustrates, once again, both the potential lines of fracture in their own ranks and their absolute unanimity in opposing the UMW. Those working their mines with free labor did not hesitate (privately, at least) to vent their frustration with the system or to complain of "how seriously [they were] handicapped on the selling end by convict labor." But they also expressed their "fear" of "offending our friends" by raising the matter in public. And lurking behind their hesitation was a concern that the UMW would be the main beneficiary of abolition.

For obvious reasons, the union had long been at the forefront of efforts to take convicts out of the mines and played a prominent role in the campaign taking shape simultaneously with DeBardeleben's effort in 1915. This fact more than any other tempered the operators' enthusiasm for attacking convict lease and in the end led DeBardeleben's aides to reconsider their position. Cyrus Garnsey Jr. of the Galloway Coal Company warned De-Bardeleben that "the movement among . . . free miners to arouse a state wide sentiment" against convict labor was "likely to be more effective than anything operators could do" and admitted that he was "more afraid of the miners' movement hurting us, than I am of the convicts. It is a tough proposition any way you look at it." G. F. Peter of the Southern Coal and Coke Company concurred, fearing that removal of the convicts from the mines would "hand over to the [free] miners the mining situation in this State, with the result that the cost of mining will probably advance so as to eliminate any advantage that we may gain on the selling price."[48]

DeBardeleben's trusted mine boss Milton Fies had been outspoken in advocating abolition early in the campaign, but Fies began to reconsider his position in light of the UMW's prominence in the movement. He thought that "the injection of [District 20 president] Kennamer and the UMW into this proposition is a serious matter and it ought to be stopped, if possible, *provided it is not entirely stopped.*" As an alternative to outright abolition, he pursued a compromise with the convict operators, initiating discussions with Pratt Consolidated's George B. McCormack and refraining from further public condemnation of convict labor. Relieved of having to conduct their defense of the system on two fronts, convict operators were thus able to weather the most serious challenge to the system in its history.[49]

From its inception, the boundaries of the operators' reform campaign were sharply circumscribed by their commitment to profitability and their stake in the ongoing exploitation of the mine workforce. These priorities, along with their determination that the project should not be allowed to escape their own direct control, diminished any humanitarian potential that the movement might have seemed to offer. There were many areas, however, where the operators' long-term interest in a stable, productive workforce and mineworkers' aspirations for a better life seemed, at least nominally, to coincide. Welfare capitalism helped to stabilize the district in the aftermath of

1908 partly because the balance of forces in the district during this period precluded a collective challenge to the operators' hegemony but also, importantly, because the system delivered sufficient material benefits to extract a certain amount of loyalty from employees.

On one level the district's transformation was impressive. The historian Marlene Rikard has argued in regard to the welfare work initiated by TCI that "the particular milieu of the Birmingham District gave rise to a much more extensive range of programs . . . than was developed at any other subsidiary or possibly at any other industrial concern in the nation." TCI's efforts were significant because they represented an ideal toward which all of the progressive companies in the district were striving—comprehensive, painstakingly planned and coordinated, and adequately capitalized. They are also instructive, however, for the relationship they reveal between compulsion and initiative in the construction of the operators' new order.[50]

The compulsory character of so much of the district's material renovation can best be seen in the evolution of TCI's first model camp at Docena. When completed, Docena was undoubtedly one of the most attractive and well-equipped mining camps in the United States. With a name derived from the Spanish word for the number twelve (*doce*), the town was attractively "laid out around a public square called the Prado. Neat four-room bungalows were provided for workers and their families. . . . A brick commissary, known as the Mercado, faced the central square," along with "other public buildings [including] Baptist and Methodist churches and a Masonic lodge." Few observers could fail to have been impressed by the contrast between conditions at Docena and those prevailing elsewhere throughout the district.[51]

Less obvious, however, was the extent to which construction of the model camp had been forced upon TCI management. Docena was, after all, the reincarnation of TCI's number 12 mine, which prior to 1912 had been worked by a mixed workforce comprised of convicts and free miners. To his credit, the company president George Crawford had solved the dilemma forced upon him by the withdrawal of convict labor by building a new camp from the ground up, but even here his hand had been forced. Free miners at the number 12 rejected Crawford's earnest appeals to accept temporary housing in an "old convict jail in the mountains near the mines," insisting instead upon standard housing. Their defiance further complicated TCI's transition to free labor, rendering it necessary for the company to undertake costly new construction immediately.[52]

The construction of model communities at Docena, Edgewater, and elsewhere provided TCI with an unmatched asset for attracting skilled labor and demanding discipline and loyalty from their employees. The "primary reason" for company housing among southern coal operators, according to the Department of Labor, was to "give stability to labor supply," and as early as 1912 observers had noted the "tendency" of the "better men" to "gravitate to the better camps." TCI set out to displace the "great floating element" of mine labor in its camps with more stable, family-centered miners, whom it supposed would be less prone to pick up and move on at the slightest offense. The company's welfare policy was therefore aimed from the beginning at attracting married men and their families and "anchoring" them to the camps.[53]

Officials considered the coarse, frontier-town quality of the district's mining camps an obstacle to securing a stable labor supply and conceived the model camps quite deliberately as havens from the rough-and-tumble atmosphere that prevailed elsewhere in the coal district. Given the scarcity of investment capital in the early years of the industry and the geographical isolation of the mines, the harshness of camp life had been to some extent inescapable before the turn of the century. In his report to the ACOA, Morris Knowles noted that mining camps were by nature "unusual communities in many ways." Located as they were in "rough, unsettled regions far from the beaten track of commerce and travel," he considered them "more isolated than any other kind of settlement." Their isolation inevitably reinforced the despotism of individual mine owners. "The land is owned by the mining company; the houses are owned by the mining company. The villages are not incorporated," John Fitch observed. "The whole thing is private property."[54]

The despotic and transient ambience of the coal camps, the "monotonous character of the work," and the fact that the population of the camps was "to an unusual extent made up of adult males" meant also that daily life in the camps could be extremely coarse. A sympathetic journalist described the "average negro mining camp" in 1907 as a "scene of squalor and desolation": "The work is hard and dangerous, and the negroes react from it into rioting and drunkenness." Among both blacks and whites, gun battles were a regular occurrence, and operators complained that the "pistol-toting evil" that had taken hold of Birmingham was rampant in the coalfields, most ominously—from their perspective—among black miners. "The miners [were] almost constantly receiving through the mails little catalogues from the big firearms companies, advertising their 'specials' at prices ranging anywhere from 98 cents on up," one source observed, speculating that "the average negro sees these 'beautiful' pistols . . . and his natural desire to walk around with a gun in his pocket prompts him to send for one of the cheap guns."[55]

The bleakness of camp life during this period brought large numbers of miners on regular sojourns into Birmingham, where their wages could buy alcohol, entertainment, and sexual companionship. Areas like "Buzzard's Roost" became notorious as miners' hangouts, giving the city the feel of a western boomtown. An early study noted that the city was "always crowded with workers wanting release from the long dreary hours in the pits. They turned to the saloons for solace, and with liquor flowing freely, houses of prostitution easily available, and gambling just up the alley there followed a natural corollary of crime and debauchery."[56]

Mine owners frequently complained that their operations suffered as a result of such unwholesome activity. ACOA officials considered the weekly sojourn into town an "important [cause] of unreliability and inefficiency in labor." At some mines, a Pratt veteran recalled, management attempted to curb the miners' proclivity for this form of release simply by paying off as late as they could on Saturdays to prevent the men from going into town. At leading operations such as TCI, however, management attempted, under the framework of company welfare, to replace such "vicious forms of relaxation" with what it considered more salubrious entertainment. Morris Knowles urged operators to secure movie projectors to "[compete] with saloons and blind tigers," and some of the major coal companies took his advice seriously, building (segregated) clubhouses and sponsoring choral groups, brass bands, and athletic teams to lure miners away from inebriation and save them from "demoralization." TCI went even further, circulating a pamphlet among its employees entitled "Teaching Thrift in Spending Wages," an action predicated, according to the *Labor Advocate,* "upon the theory that if men can be taught to live more economically they can be forced to work for lower wages."[57]

At the center of TCI's blueprint for camp reform was the cultivation of a stable domestic life for employees and their families. In this endeavor they devoted considerable attention to winning the loyalty of mining camp women. The average miner spent so much of his time underground, one miner's wife complained, that "if it was cloudy he didn't [ever] see the sunshine . . . and his kids didn't hardly know him." The children were "asleep when he left and asleep when he come." Women therefore carried much of the day-to-day burden of holding families together, a role that included not only child-rearing but virtually every task other than actual labor in the mines. When her husband was in the mines, Ila Hendrix recalled, "I worked everywhere else." Like other mining camp women, Hendrix managed a small family garden and raised cows and chickens, fashioning dresses out of leftover cow-feed sacks for her young daughters. Annie Sokira, an immigrant and mother of ten, recalled the desperation that gave added significance to management gestures for camp women. She remembered trying to send her children to school during strikes "'til sometimes you'd patch and patch 'til there wasn't nothing but a patch and the patch wouldn't hold the patch no more."[58]

"The solidarity of miners underground," David Montgomery has written regarding American coal unionism generally, "was reinforced powerfully by that of women aboveground."[59] The prominent civic roles assumed by camp women, along with the operators' assumption that experienced, dependable miners would be more willing to sink roots in those camps offering decent amenities for their wives and children, fixed the attention of company-paid "welfare workers" on winning their loyalty to the employers and shoring up stability in the miners' homes. "A man's family must be satisfied as well as the man," Ben Roden explained to a group of fellow operators in 1911, "else he keeps moving from camp to camp . . . [costing the operator] from at least three days to as many weeks loss of that man's coal output, until he is re-placed." A female guest speaker at Henry F. DeBardeleben's Margaret mine told the company-sponsored welfare association that "the men of any community are just what the women make them." "If a man is prosperous," she explained to a captive audience of miners' wives, "some good woman is push-ing him. If a man is shiftless and fails to give his best service, some woman is the cause of it. She fails to prepare his meals on time, or leads him into some unprofitable field." TCI's Dr. Lloyd Nolan, appointed to direct the company's ambitious health and sanitary efforts, reminded mine owners that "the sober, industrious and hardy workman must have properly prepared food, a clean and well-conducted home and healthy family." It followed, he warned, "that wives and prospective wives who are ignorant of housekeeping, sew-ing, cooking and of home hygiene, must be instructed if efficient workmen are to be produced." Nolan's protégé, Marion Whidden, depicted the typi-cal mining camp as "a collection of families" with problems very much the same as those found "in any small city or town."[60]

Model mining communities were thus conceived as something more than mere physical upgrades of deteriorating housing stock. They were intended to serve as the anchors of a new social compact in which the mutual needs of mine owners and their employees could be reconciled and the antagonisms common to modern industrial life purged from camp life. To succeed, the project required not only material resources but also moral instruction un-der the leadership of the employers. Although architects of the new order viewed themselves as "progressives" with the best interests of their own workers at heart, the assumption that progress would proceed under the direction of and at the pace prescribed by the operators was implicit in its design. The material benefits bestowed upon the mine workforce came at the expense of the "miners' freedom": "Granting workers the privilege of living in Docena gave TCI an important means of social control," Lewis observes. "Mine superintendents could force residents whose behavior did not measure up to company standards to leave after a single day's notice" while "company-sponsored welfare workers and medical practitioners swarmed throughout the community conducting home visitations, making sure that children brushed their teeth, seeing that they changed their underwear . . . such paternalism was the price that workers had to pay for the improved conditions they were allowed to enjoy."[61]

In the TCI camps, workers troubled by such intrusions into their personal lives could still point with some satisfaction to the material benefits they derived from employment with the most advanced company in the district. On one level, TCI appeared to treat the challenge of anticipating employees' grievances seriously. They removed one of the most serious causes of dissatisfaction by adopting twice-monthly paydays in place of the traditional monthly system which, critics charged, made employees dependent on high-interest advances from company commissaries. Management introduced uniform, regulated pricing at its camp commissaries at a time when some of the smaller coal companies were only half-jokingly referred to as "grocery companies that operated a few coal mines in order to create a captive client of miners." TCI's extensive medical program, which eventually included the construction of a fully equipped, modern hospital at Fairview, had earned the company national acclaim. And the educational facilities it sponsored were without equal in the mining district. "In areas where municipal, country or state educational facilities [were] poor," one company historian recounted somewhat exaggeratedly years later, TCI "gladly assumed the burden." The willingness of county authorities to "turn over to the company the annual appropriations received from the State for teachers' salaries" left TCI in complete control of education in the camps. Teaching candidates, like other company officials, were closely vetted to ensure their compatibility with company interests, and miners' children were offered a curriculum "free from the modern 'Isms'" that operators feared would lead to "revolution and ruin." Despite these shortcomings, given the desperate state of public education throughout Alabama, many employees—and perhaps blacks more than whites—valued the opportunity to send their children to functioning schools.[62]

While leading employers could afford a solution to the problems of labor supply and stability that offered tangible material rewards to miners and their families, elsewhere the coercive aspect of company paternalism was far more pronounced. The real achievements associated with the new order filled the columns of the booster press and won the admiration of reformers and industrialists far beyond the Birmingham district. But the operators' grip over the district was apparently never secure enough to allow the relaxing of their vigilance or the dismantling of their formidable police apparatus. The "spirit [of] cooperation" that John Fitch had found so "tremendously lacking" in his early evaluation of company welfare never materialized, and compulsion continued to play an important role in maintaining the system's viability. Despite his admiration for the work being carried out by TCI, Fitch recalled that during his travels around the district he had been warned to "be careful about [trespassing upon] certain Tennessee camps," that company deputies would not hesitate to use violence against him. After the dust from the 1908 strike had settled and the UMW threat had receded, coercion continued, but in a milder form. When in 1913 the federal government began its investigation of U.S. Steel, TCI management posted petitions supporting the company at mine tipples, but "very few [employees] signed them" until mine foremen were instructed to approach workers individually to "request" that they sign. On another occasion, when a TCI foreman at Docena "went through the mine soliciting donations for the Red Cross," he approached a black miner and suggested that "$5 was what he should contribute," offering to spread the necessary payroll deductions over a four-month period. Since, according to the *Labor Advocate,* "a suggestion from the foreman was equivalent to a demand," the "negro agreed to pay." When several days later the same miner was discharged for membership in the UMW, he found that the entire amount had been deducted from the wages owed him.[63]

In the absence of a formal, union-enforced grievance system, operators developed procedures aimed at providing workers with a tolerable means of expressing their grievances. At many smaller, isolated mines this involved no more than a pledge from the mine boss that his office was "always open to the lowest-paid worker in the mines." At larger operations, management adopted more elaborate measures. Separate black and white "welfare committees" held weekly meetings at the Alabama Fuel and Iron Company, but they represented little more than a mechanism for inculcating company loyalty. "Local superintendents are present," an internal report assured stockholders, "for the purpose of encouraging . . . employees to cooperate with the company to improve their conditions of living and labor." TCI placed "suggestion boxes" at their mines and established a "Mutuality Department" as a channel for pressing workers' complaints. For obvious reasons, however, employees placed little faith in the impartiality of such institutions. Such committees, the UMW organizer William Harrison complained, "like the elections of checkweighmen, are . . . under the influence and domination of the Company and their agents. Consequently . . . [they do] not and can not afford any protection or benefit to the Miners." "Woe to the man who takes his case to the Mutuality Department," one disgruntled former TCI manager warned. "They are always put on a black list in the employment office marked 'NG' [no good]."[64]

The ACOA encouraged its members to implement grievance structures but supplemented this initiative with an elaborate and well-funded surveillance program. The result, according to UMW officials, was that there were "only a few places where we can hold meetings." Although union officials in 1913 reported that "sentiment for organization could not be any stronger . . . than it is now," miners were "afraid to attend" meetings, J. R. Kennamer reported, "because there is always someone spying." Reports surfaced in 1913 that the operators were discharging disgruntled miners by sending "fake [union] organizers" through the district in an effort to entrap them. Management at Sipsey "question[ed] all who desire[d] employment . . . very closely, to avoid hiring any men with union sympathies," and when Milton Fies suspected one of his section foremen of harboring UMW sympathies, he immediately fired the individual and imported ACOA detectives to "make a thorough investigation as to just how much trouble" the man had "caused." Some years later, when men in the company's employ at Payne's Bend stopped work to protest a missed payday, company officials instructed supervisors to "find out who is at the bottom of this movement and discharge him."[65]

"There is no power to require the United States Steel Company to be fair to its employees," Fitch testified in 1912, "other than the willingness of the corporation to be fair, or the power of the workmen to combine and tie up the industry through strikes." The latter possibility had been at least temporarily ruled out by the operators' victory in 1908. For nearly a decade afterward, district coal miners were at the mercy of their employers. The viability of welfare capitalism demanded first and foremost that the miners remain divided and unorganized. But the seeds of its own destruction were present in the new order from the beginning.[66]

Unquestionably, the operators' initiatives had forced coal unionism onto the defensive—so much so that in 1914 the UMW proposed its own version of social welfare, nearly identical to the plan envisioned by mine owners with the single exception that the work would be carried out "under the jurisdiction of the UMWA."[67] But for all its apparent impregnability, the operators' new order was vulnerable to attack. Surveying the record of employer paternalism in southern mill villages, Broadus Mitchell and George Sinclair Mitchell concluded that the system had the effect of rendering textile workers "uncomplaining and dependent." "The operatives in company towns," they wrote, "have been stall fed." That was certainly one potential effect of welfare capitalism in the Birmingham district coal mines. In their classic survey of black workers' relation to organized labor, Spero and Harris conclude that TCI's welfare policies "gave the company a working force in which trade unionism could make little headway." "Even in the mines," they report, "where there was a long tradition of militant labor activity, the union was not successful." Their observations are confirmed by the remarks of a frustrated union militant from Blocton, who complained that there were too many miners in the district "who would rather have a smile from the boss than a ten-dollar bill from some honest source."[68]

But welfare capitalism also had the potential of backfiring on its overseers—of compounding miners' grievances rather than dissolving them. By 1915 that is precisely what had begun to develop in the mining camps: deep resentment over the authoritarian character of welfare capitalism and a growing desire to burst the walls of the "stall." "There ain't no such animal as 'uplift work,'" a correspondent wrote in the columns of the *Labor Advocate*. "If labor were to receive its just share of the profits on what it produces, there would be no necessity for nine-tenths of the charity work that is done." Despite tangible improvements to "general working conditions," Marlene Rikard has written, "many of the workers' grievances went unheeded" until after the First World War. Many operators misread the apparent calm that prevailed in the coalfields after 1908 as evidence that their employees were not merely satisfied but pleased with the new arrangement. When the convulsions brought on by World War I opened up new possibilities for union organization, mine owners would be shocked and exasperated at the distinct lack of gratitude and the fury and determination that workers exhibited.[69]

3. "Friends" in High Places:
Racial Paternalism and the Black Miner

THE TRULY DISTINCTIVE FEATURE of welfare work in the Birmingham district was the operators' systematic attempt to integrate the region's deeply rooted tradition of racial paternalism into the edifice of modern industrial relations. In national terms, the basic features of company welfare in the Birmingham district seem unexceptional. The South's legacy of late industrialization and harsh regional poverty meant that leading companies began their work of reconstruction on a lower material level than many of their counterparts in northern industry. But the general formula adopted by district operators—increased productivity and labor peace through moderate, targeted investment in the health and well-being of their employees—had by 1910 been adopted as a standard recipe among many leading northern employers. Neither the scale of the operators' undertaking nor the receptivity of district miners to their initiative distinguished the Birmingham variant of welfare capitalism. Instead, as Horace Mann Bond insisted, it was "the peculiar racial situation of Alabama workers" that "permitted the development of a paternalism unmatched anywhere else in the country."[1]

The failure of the employers' immigration campaign marked an important point of departure for the district. Above all, it reoriented the district's major employers to the exploitation of indigenous black and white labor. Mine operators had certainly recognized the advantages offered them by black workers' vulnerability under Jim Crow since well before the turn of the century. Their frank realization that neither native whites nor immigrants would willingly labor for the paltry wages or under the dismal conditions that they imposed on blacks tempered the operators' frustrations with their alleged inefficiency. With slight alteration, industrial Birmingham's experience

By the turn of the century, African Americans made up a majority of the Birmingham district mine workforce and nearly three-quarters of all mineworkers by the end of World War I. Aided by local race leaders, coal operators cultivated close ties with what John Fitch referred to in his January 6, 1912, exposé in *The Survey* as their allegedly "cheap, docile negro labor" in hopes that black workers would frustrate attempts at union organization. (Courtesy of the Robert D. Farber University Archives, Brandeis University, Waltham, Mass.)

mirrored that of an Alabama planter who pleaded the superiority of black field workers by reminding his peers that "no other laborer . . . would be as cheerful, or so contented on four pounds of meat and a peck of meal a week, in a little log cabin, with cracks in it large enough to afford free passage to a large cat."[2]

The formidable political, cultural, and legal foundations upon which Jim Crow rested reinforced the ability of planters and industrialists to set the terms under which their overwhelmingly black labor reserve would be employed. While the arrangement offered little to inspire cheerfulness among those pushed to the bottom of southern society, it devoted considerable resources to ensuring that their frustrations would never be publicly aired or effectively redressed. Black workers' precarious standing in a region committed to the strict enforcement of white supremacy made it hazardous—and unhealthy—to complain and next to impossible to mount coordinated, collective resistance in the form of work stoppages, boycotts, or public demonstrations. These restrictions left black southerners perilously suspended between slavery and free labor, and employers in both factory and field were alert to the substantial advantages afforded them by this state of affairs.

The "preference" for black labor so frequently articulated by Birmingham coal operators derived from their recognition that, with black workers deprived of any rights that white capital might be obliged to respect, they could be more easily exploited than white miners. The operators' outlook paralleled that of their counterparts in the Black Belt, who complained that the problem with white tenants was that "they want more advances and you can't hold them down the way you can a Negro." "If you tell a Negro he can't have any more, he will go back to work," one planter remonstrated, "but a white will grumble and won't work, and will even move out on you." While the legal restrictions attending the caste system provided an effective means for disciplining black laborers, planters objected that white tenants enjoyed an "ability . . . to resort to legal defense against dishonest settlement, terrorization, illegal eviction or illegal seizure of livestock and personal property." "Give me the nigger every time," a Mississippian pleaded. "He will live on less and do more hard work, when properly managed, than any other race or class of people."[3]

Not only in Black Belt agriculture but in industrial settings where black workers were employed in large numbers employers strove to refine a systematic approach to the "proper" management of black labor. In a period when the status of black workers hovered somewhere in the "twilight zone between slavery and freedom," the authoritarian ambience of the coal camps assured the easy transfer of plantation-style discipline to Birmingham's mineral district. Before the turn of the century, the direct involvement of individual planters in mine management facilitated this transit. At least one planter-cum-operator, the Virginia transplant Joseph Bryan, "habitually used plantation analogies in discussing industrial problems with his associates and saw nothing unusual about entrusting day-to-day control of his enterprises to administrators drawn from . . . the southern planter class."[4]

Well into the twentieth century, district employers continued to place a premium on their foremen's aptitude for "handling" black labor. "The 'boss' who can get the best work from a crew of southern darkies," one district veteran explained, "must be a man of unusual gifts." Lauding the success of an "ideal southern mine boss" in "making his gang of five hundred negroes as efficient as any equal number of whites could be," an *Age-Herald* correspondent paid homage to the Sloss superintendent "Captain" John Hanby: "Generous, bluff, convivial, one minute knocking a negro down for disobedience and the next minute picking him up," Hanby's admirer reasoned that the district's "labor problem" could be easily solved if only mine bosses throughout the coalfields would emulate "Cap" Hanby's "rough and ready" style. Decades of such experience led operators to the conclusion, here articulated by a leading advocate of "progressive" management, that "there is no superior to the negro as a willing, loyal, reasonable, and obedient laborer." Their singular defenselessness—the lack of recourse available to black miners who might object to the treatment meted out to them—convinced operators that black labor was not only "an economic asset" but "an absolute necessity for the industrial consummation of industrial Alabama."[5]

The viability of paternalism rested, therefore, upon the stifling subjugation of black labor. When circumstances demanded it, the system's beneficiaries would not hesitate to assert their authority by applying the harsh methods endorsed by the *Age-Herald*. But in the years following Redemption, southern employers enjoyed such a solid hold over their workforce that less abrasive methods were often sufficient. The operators' seeming omnipotence guaranteed the continuity of yet another antebellum tradition: black laborers' personal dependence on the goodwill of individual employers. In paying "just tribute to the worthy Southern negro" before an audience of northern manufacturers in 1916, the Kentucky industrialist Frank D. Rash lauded black labor as "one of the greatest assets of Southern industry." Wielding an embellished version of the late Booker T. Washington's prescription for racial advance through "work, hard work, and efficient work in the fields, the factories, the forests and the mines," Rash noted the survival among black laborers of that "now too rare feeling of deep interest and loyalty to [one's] employer." "When the Southern Negro works for a corporation he calls it 'my company'; his loyalty becomes deep and unchanging and the sower of seeds of discord finds little response to his beguiling . . . stories of greener fields."[6]

While Rash's paean to Negro loyalty was rooted in traditional, racist assumptions about black docility and conveniently ignored the circumstances that fastened this quality upon black labor, he articulated a perspective common to many men of his class and one that contained an important kernel of truth. "The average Negro is more accustomed to work for persons than for wages," Booker T. Washington had acknowledged in 1913. "In the past he has found that the friendship and confidence of a good white man, who stands well in the community, are a valuable asset in times of trouble." Not only accommodators like Washington and those who, like Rash, stood to profit from submission but even individuals attempting to rouse black laborers to self-defense acknowledged the formidable staying power of antebellum servility. "That relic of slavery is still left," Eugene Jones, an NAACP official, observed several years later, "which causes the Negro to believe that he should be slavishly loyal to the man who employs him." This he attributed to their having "been taught for three hundred years the individual relationship between employers and employee. . . . They speak of 'my' bank when they are only the messengers in the institution." It followed, for Jones, that any attempt to ease black workers' predicament would inevitably fall short unless it succeeded in convincing them to reject "that old ante-bellum theory of absolute dependency upon the man for whom [they are] working from day to day."[7]

District operators seemed determined to prevent such a development. Instead they moved to take advantage of black workers' predicament, hoping that the skillful cultivation of their loyalty would provide a powerful bulwark against labor unrest. Even in its most benign apparition, coalfield paternalism bore the unmistakable signs of its lineage in antebellum plantation agriculture. Speaking to employees at a company-sponsored outing at Aldridge mines in 1917, the owner A. B. Aldridge revealed that it was "no new thing for his company to . . . give their employees a barbecue and picnic." He insisted that he was "only following . . . the old plan of my parents of nearly five generations," whose "custom" it had been "to give the employees an old-fashioned outing once every year." A former employee of "Charlie" DeBardeleben's Alabama Fuel and Iron Company captured vividly the antics that formed a routine element in camp life at Margaret. "He'd give a big barbecue and all that type of stuff and he'd make his speech and ask . . . , 'Who's the greatest man in the world?' And [the miners would] get to clapping, 'Uncle Charlie DeBardeleben, Uncle Charlie.'"[8]

These relatively innocuous displays of paternalism were complemented, however, by more serious and sinister measures aimed at preventing any challenge to the operators' authority. The same miner who mocked the crudeness of DeBardeleben's methods also acknowledged that he was "the toughest" of the antiunion operators and described Margaret as a "regular slave camp." Years before their formal embrace of welfare capitalism operators had recognized the strategic importance of black mine labor as a barrier to unionism. In doing so they illustrated the "disposition" among southern employers—noted by an English visitor to the region after the overthrow of Reconstruction—"'to rely on black labor as a conservative element, securing them against the dangers and difficulties which they see arising from the combinations and violence of the white laborers in some of the Northern States.'"[9]

The best example of these early efforts was the proclamation issued to black miners by "Uncle Charlie's" father, the TCI official Henry F. DeBardeleben, during a strike in 1894. Frustrated by union efforts to organize the company's Blue Creek mines, DeBardeleben offered to establish an all-black scab colony there in return for a guarantee of labor peace. "A job at Blue Creek is a desirable one," he advised black miners in an appeal aimed explicitly at eliciting their support in breaking the strike. "[Blacks] can have their own churches, schools and societies, and conduct their social affairs in a manner to suit themselves, and there need be no conflict between the races." While such an appeal understandably found a resonance among black miners in a period of hardening racial lines and intense racial violence, DeBardeleben went a step further. "Colored miners, let us see whether you can have an Eden of your own or not. I will see to it that you have a fair show. You can then prove whether there is intelligence enough among colored people to manage their own social and domestic affairs."[10]

DeBardeleben's appeal prefigured the operators' approach to black labor after 1908. While borrowing heavily from old-style paternalism (the pledge to act as guarantor of a "fair show" for blacks and to ward off the "interference of the white race"), his cynical attempt at manipulating black racial pride was the most striking aspect of DeBardeleben's appeal. In a period when Alabama blacks saw their rights in the political sphere increasingly circumscribed (disfranchisement would come in 1901) and when extralegal violence commonly reinforced legislative restrictions, DeBardeleben's "Negro Eden" held out to black miners a measure of autonomy and racial self-government. Deliberately wrapped in a challenge to racial self-esteem, the proposition represents an attempt to put the best face on the newly ascendant doctrine of "separate but equal" and harness it to the benefit of the coal operators; to hold out the semblance of power now that black self-determination had been shorn of any real substance.

The lack of a viable, pragmatic alternative rather than any innate attachment black workers felt toward their employers ensured the continued survival of the paternalist arrangement. The social predicament of black miners made them somewhat of an unknown quantity in labor disputes: depending on a number of contingent factors (including, importantly, the attitude of white workers), under different circumstances they could exhibit either bold, audacious militancy or fearful submission. Certainly any experienced coal operator knew firsthand that their public claims of eternal submissiveness and innate black hostility to trade unionism were dubious. As every major confrontation in the coalfields had demonstrated, wherever circumstances made effective, mass industrial action possible and rendered untenable the operators' strategy of individual victimization black miners were often the most militant defenders of the union. And during the bleak years following defeat a core of experienced black unionists played an important role in maintaining a UMW presence and staving off the union's complete demise.

While the evidence refutes the operators' contention that black miners were racially predisposed to docility and antiunionism, there is no doubt that many aspects of their day-to-day lives had fitted Alabama blacks for their designated role as a bulwark against union organization. The sheer poverty of the Cotton Belt brought a continuous stream of black laborers north to the mineral district in search of work, and the reaction of one new arrival as he stepped off the train at Birmingham, that the city "looked like heaven to me," was not atypical. One miner, F. C. Jones, recalled that he came to Birmingham because "she was a Magic city, Magic city." Although newcomers were generally not long in discovering just how far from heaven they had landed, naturally it required time before their euphoria gave way to despair over having traded rural poverty for the industrial squalor of the steel mills and coal camps. Although his landlord had "begged [him] to stay," John Garner made Birmingham home. "If you told [industrial employers] you come off the farm, they'd hire you in a minute cuz they knowed you didn't know nothing but work." Meanwhile the destitution that awaited those who harbored thoughts of a return to the Black Belt played an important role in reconciling them to conditions they had not anticipated. The experience of Will Armstead, a black miner, is typical. Determined to leave behind the "never-ending cycle of debt which ensnared sharecroppers" in the Cotton Belt, Armstead literally "[snuck] away" to Birmingham and landed a job in the Mulga mine where, along with wife and family, he was allotted company housing. As Ronald Lewis has noted, Armstead's new life seemed "a significant step up in the world," and "it would take some time for him to fully comprehend the new relationship of dependency he had assumed" and begin to consider open resistance to the operators' power. "In the meantime," Lewis writes, "company paternalism meant that he had exchanged a malevolent master for a benevolent one, and it would be an act of gross ingratitude to turn on such an employer."[11]

Only when compared to the precariousness of sharecropping and the stifling monotony of plantation life did life in the coal camps seemed superior, and then only until new arrivals had experienced enough of camp life to appreciate how little their lives had actually changed. The continuous influx of Black Belt refugees meant that, just as they became adept at employing the threat of cheap black labor generally to contain the demands of white miners, operators found they could use the threat of displacement by new arrivals to temper the dissatisfaction of those black miners who had grown restless over conditions. During periods of labor agitation, for example, UMW organizers reported little trouble winning experienced black miners to the union and even succeeded in convincing large numbers of new arrivals to spurn their role as strikebreakers. "Labor agents are bringing negroes into this vicinity in great numbers claiming to be building a railroad," strikers at Jasper would report during a later strike. "All we are able to get in touch with are more than anxious to leave when they find out they are going to work in a mine, and all they ask is railroad fare away from here to Birmingham." "They gets a lot of colored fellows in here but they don't stay long," reported another from Parrish. But invariably such efforts were overwhelmed by the arrival of trainload after trainload of strikebreakers fresh from the Black Belt.[12]

Black miners' predicament was further complicated and the appeal of racial paternalism elevated by the pervasiveness of white hostility. DeBardeleben's Blue Creek Proclamation ushered in a halting but clearly perceptible drift toward a "deliberate policy of flattering the Negro workers" and "exalting [them] as competitors of the whites," and such a strategy resonated among black miners in proportion to the antipathy they faced from their white co-workers. While organized labor had never exhibited a monolithic attitude toward Birmingham's black working class and by some accounts represented "one of the few glimmers of racial liberalism" in the state, relations between blacks and whites in the coal camps—particularly after the 1908 defeat—no doubt reflected the resurgence of popular racism in the years following Redemption.[13]

It is not surprising, therefore, that black miners occasionally failed to distinguish between the relatively egalitarian policy of the UMW and that of the labor movement generally, which "was not only neglecting to organize Negro workers, but . . . was following a deliberate policy of exclusion." The dilemma facing black workers throughout the United States was cogently summarized by the *Indianapolis News* after a deadly local confrontation between black strikebreakers and mostly white union miners: "If non-union men are not permitted to work and colored men are not permitted to join the union, where does the colored man come in? Or does he stay out?" The answer provided by DeBardeleben and later adopted by operators throughout the Alabama fields was that the black man "comes in" under the benevolent protection of his employer. And to the extent that the UMW fell short of establishing equality in its own ranks, the operators' formula inevitably won supporters among black miners.[14]

The black miner stood to benefit in small but significant ways by maintaining amicable relations with his employer, whose influence could be valuable in warding off racial transgressions at work, for example, or for keeping black workers out of the clutches of overzealous white policemen. In both cases operators stood to benefit as much as the employees they protected. When a storekeeper's nephew assaulted the wife of a black miner at Sipsey, the company president Henry T. DeBardeleben ordered the boy out of his camp, fearing that "occurrences of this kind will influence greatly at our cost the negro labor." The business of intervening to spare a loyal employee jail time seems to have been lucrative, as operators typically charged substantial interest on the amount posted for bond. But there were other important considerations at play also: in its fervent defense of the racial status quo, the southern legal system seldom differentiated between unrepentant lawbreakers and upstanding citizens, and operators worried that the experience of sharing a cell with hardened criminals might "demoralize" otherwise conscientious laborers. One came to the aid of one of his "good servants" in order to prevent him from "mingl[ing] with convicts," an experience that he feared might transform his employee into a "bad Negro." Mine owners did not merely intervene to spare their employees the ravages of racial injustice, however; they sought to reward loyalty with privileges that were otherwise rarely available to black workers. A skilled black employee who had proven his fidelity over an extended period of service and who had little chance of securing a bank loan elsewhere might approach his mine superintendent for an advance. Gestures of this kind, which cost the operators little or nothing in material terms, went a long way toward securing the goodwill of black miners.[15]

The viability of racial paternalism did not depend, however, upon the approval of rank-and-file black miners. Unilaterally imposed, it was not, after all, the result of a negotiation among equals but of an initiative emanating from the employers. To the extent that they solicited the consent of blacks at all, it was Birmingham's small but influential black middle class and not black miners that the coal operators looked to. In spite of the considerable advantages they enjoyed, the operators' system of racial paternalism never succeeded in securing the allegiance of the mass of black miners. Its most outspoken proponents among blacks were not the miners themselves but black elites—preachers, businessmen, newspaper editors, camp welfare workers, leaders of the fraternal orders—who were held up by employers as the natural spokesmen for black racial progress, and who were often materially supported by the coal operators. The operators' cultivation of strategic ties to Birmingham's conservative, accommodationist black leadership aimed at developing a crucial strategic asset for the construction of racial paternalism in the coalfields.

The unique constellation of social and economic forces at work in the Birmingham district after 1908 gave rise not only to an ambitious attempt to establish the foundations of welfare capitalism but to a variant of that system that reflected and in many ways reinforced the racial divisions extant throughout southern society. To the extent that the arrangement inspired loyalty among miners of either race, it did so in part because it delivered real improvements in the cultural and material conditions under which black and white workers lived. But those companies in the forefront of the welfare experiment very definitely did not, as one charmed observer of TCI's work supposed, "disregard racial lines" in their operations. On the contrary, the deliberate manipulation of racial divisions remained central to their efforts to fix the atomization of the industrial workforce as a permanent feature of labor relations. And despite their promises of an earthly Eden, the operators' paternalist edifice stood squarely on top of a system of harsh exploitation—one that systematically doled out brutality to black miners.[16]

No feature of life in the coal camps better epitomized the contradiction between the bombast and substance of paternalism than the notorious convict lease system, and none better illustrated the continuing centrality of racial subordination to coal industry profits. Convict labor was not a peripheral feature of the Alabama coal industry but instead played a fundamental role in setting the conditions under which the district's free miners labored, functioning simultaneously as "an agency for racial control and supervision." Lewis reports that throughout the period between 1900 and 1928, when the system was abolished, blacks made up over 80 percent of all state convicts, and their representation among county convicts was similar. Indeed, a high percentage of black miners, perhaps even a majority, dug their first coal while decked out in a prison uniform. "The occupation of mining has been opened up to the negro," a warden at the state-run coal mines in Tennessee observed, "although his entry into the craft has been principally through the rugged gates of the penitentiary."[17]

The coal operators' insatiable demand for cheap, tractable labor, the sheriffs' substantial material stake in zealous enforcement of the law, and the ubiquity of antiblack racism contrived to usher the heavily burdened black laborer through the penitentiary gates. The overwhelming majority of those sentenced to mine labor were arrested on trivial charges like gaming, vagrancy, or petty theft, and the rigor with which these laws were enforced fluctuated according to changes in labor demand. The system thus accomplished two central objectives for the New South regime: it offered an effective instrument for disciplining black southerners and reinforcing white supremacy while securing a steady flow of cheap, unfree labor for the mines. As Alex

Lichtenstein has argued, the "development of the Lower South's coal and iron industry . . . rested on the ability of southern capitalists to use the penal system to recruit the core of their productive labor force."[18]

The treatment meted out to those incarcerated in the convict camps varied, as did the treatment of miners generally, according to a number of factors over which they enjoyed little control. Under the supervision of a thoughtful foreman, more fortunate convicts might avoid physical mistreatment and even acquire skills that would serve them well upon their release. Overall, however, the system proved a breeding ground for violence and racial depravity. Corporal punishment, from which even free miners were not immune, became a routine feature of life in Alabama's convict camps—so much so that the issuing of whipping licenses to convict foreman was the best that reform-minded state officials could muster to counter its brutality.[19]

The Jefferson County physician Dr. W. G. Ward's inspection of the whipping log at Sloss-Sheffield's Flat Top mine in 1906 led him to the conclusion that county convicts were being "whipped on the slightest provocation." One convict complained in a letter to state officials that authorities "treet[ed] negros like thay was hogs in a pin." Deprived of the right to refuse work under unsafe conditions, convict miners suffered an unusually high mortality rate: 90 percent of all crippling accidents and nearly all deaths among convicts occurred during sentences in the mines. Some of these, it is clear, occurred at the hands of depraved foremen. "Back then, they wasn't no pamperin'," the retired convict boss Jim Lauderdale explained to interviewers in 1934. "I remember they told us to shoot quick as hell if anybody got rough, or tried runnin' off. They said they was lots more whar these come from, an' that when you knocked one of 'em off it was no worse'n killin' a hog or a cow." Lauderdale's recollection of his career as a convict foreman at Birmingham's Red Diamond mine provides a harrowing glimpse into the brutality of the system. "We done some things that wasn't right," he recounted. "We useta keep a big barr'l out back of a shed at the mines, an' when I think back on it now, I know we whooped niggers jes' to have fun. . . . Some of 'em would pass out like a light, but they'd put up a awful howl, beggin' us to stop."[20]

The geographical isolation of Alabama's convict camps bolstered the impression among foremen that they could administer such violence with impunity and were "perfectly safe from outside interference." Following a visit to one of the convict turpentine camps in 1913, the state physician inspector J. M. Austin protested to Hartwell Douglass, the president of the State Convict Bureau, that the arrangement afforded "those who are so degraded, so inhuman . . . and so lacking in human kindness . . . an open field to satisfy their inhuman ideas about the convict and the negro." When state officials conducted an inquiry into allegations of abuse two months later, they suspected that convicts, none of whom alleged mistreatment, were withholding testimony to avoid management retribution. Upon further investigation, officials learned that "whenever an inspector comes around and any of the prisoners tell him anything that is not satisfactory, he will be whipped for it." One convict, Will Dickson, testified that his reluctance to confide in state officials was due to the fact that bosses had "whipped [him] twenty-five licks for telling things [during an earlier investigation] and I was afraid of being whipped again."[21]

A report in the *Birmingham Labor Advocate,* recounting the fate of a county convict named Charles Ford, who had been "whipped to death at the Pratt Mines" in 1907, quoted a mine superintendent's remarks that "'the company had only two places for a convict, one in the mines and the other in the hospital.'" Judging by contemporary accounts, convicts were admitted to hospitals for extended care only in extremely exceptional circumstances. Partly this was due to the poor state of camp medical care: "medical service was often inefficient [and] hospital accommodations were absent or very poor," one study concluded. More importantly, though, where foremen's wages were pinned to the coal tonnage dug by men under their authority economic incentive prodded them to "get as much work as possible out of the men, and in many prison camps it was a common thing for prisoners to receive corporal punishment for 'failure to perform work.'" Based on his experience at Red Diamond, Lauderdale reported that the companies had "lots of ways to make a bad convict work," including, in addition to physical beatings, the "sweat box" and a "bread an' water" diet. The *Labor Advocate,* a consistent opponent of the lease system, reported in 1915 that by such methods convict miners were "being driven to do more superhuman labor, to do more in one day than two men should rightfully do in two days." While some officials estimated that convicts averaged between nine and twelve hours of labor daily, J. M. Austin, after a visit to the Pratt mines in 1913, protested against convicts there being worked "from six in the morning until ten at night."[22]

Convicts reacted against such miserable conditions with acts of individual and collective resistance. Pressed to meet demanding quotas, they responded—like many free miners—by mixing slate into the coal they loaded. Management also complained that they "intentionally [broke mine] machinery." On rare occasions convicts succeeded in registering their dissatisfaction through more spectacular actions. Austin reported to state officials in 1913 that convicts at Banner were so "disgruntled and sullen" that they had "gone so far as to post notices in the mine that unless certain improvements are made, they will strike, or destroy property." Flat Top mine was the scene of a convict riot in August 1916. "The convicts have been treated so badly," a correspondent to the *Labor Advocate* reported, "that they have refused to work on several occasions; threatened to kill the guards or anyone who would come near them in their barricade . . . until the Governor or Convict Inspector Kyser came to them." And when in 1918 convicts at Banner "secured two or three Mining Batteries and fixed up some Mines with dynamite across the slope," vowing to "set off the Mine if any of the bosses started in," the warden, J. P. Hall, begged the State Convict Board not to discharge him. "You Gentlemen . . . know," he wrote, tellingly, "that convicts goes on strikes . . . at different Camps as well as free labor on the outside and there is always two sides to any question."[23]

More common than organized resistance, in which convicts confronted their jailers directly, however, was violence born of despair, in which they squandered their lives in hopeless attempts at escape or took out their frustrations upon fellow inmates. Between 1911 and 1918, a study of the state convict system reported, more than twice as many convicts died at the hands of their fellow inmates as by the actions of prison or mine officials. In May 1910 the *Labor Advocate* reported that "for the third time in a month convicts have had a fight to the death" at TCI's number 12 mine, and during the same month twenty-eight state convicts were burned alive when three of them set fire to the stockade at Lucille in a failed escape bid. Short of death, officials reported that a high proportion of convicts—"not less than twenty-five percent," according to one source—were subject to sexual assault[24] in the camps. Reformers worried that the victims, "usually young boys or men of feeble mind," would carry with them "the monstrous habits which they have acquired . . . and spread them in the outside community."[25]

The absence of more promising options available to black convicts upon their release is confirmed by the fact that so many opted to remain in the mines after the completion of their sentences. Typically, convicts who managed to complete their sentences without losing the good favor of their employers were offered continued employment and given a small stipend, along with work tools and clothing, as incentive to remain. In remarks before a federal investigative committee in 1885, a state warden testified that of the convicts discharged at the Pratt mines, "nearly all have staid [*sic*] in the mine; hardly any of them have gone away." The chief reason for this, he asserted, was that black miners could earn far more digging coal than they might receive on a plantation. Another government commission in 1912 found that the convict lease system produced "a steadily increasing supply of efficient, steady, and trained negro miners" who had not only acquired the necessary

work skills while wards of the state but, "owing to the system of rigid discipline and enforced regularity of work, [had become] through habit . . . steady
work[men], accustomed to regular hours." Their apprenticeship in the convict lease thus prepared large numbers of black miners for the slightly less
harsh regime they would endure as free miners.[26]

Aside from strikebreaking and the convict lease system, black miners' other
main point of entry into the district mines seems to have been the notorious subcontracting system, a scheme that kept wages low and complemented the operators' efforts to forestall union organization. Under this arrangement, which the Birmingham muckraker Ethel Armes insisted was of recent
origin in the Alabama fields, an individual ("not necessarily a miner, but a
butcher, a baker, farmer, or grocer") holding a contract from an operator
would "employ from four to five to as many as twenty-five laborers, for the
lowest wage they will work for." Among the crew, one or more skilled men
might receive the regular tonnage rate paid by the company while the others, paid according to a task system, would "get much less."[27]

Not surprisingly, blacks fared worse than whites under the contracting
system. Though a small number of them had, by 1907, "risen to be sub-contractors . . . having from eight to sixteen negroes under them," the more typical arrangement included a white farmer who "brought his hired hands
along—negroes who worked for what he elected to pay" or "the poorer class
of whites—for only the poorer class of workingmen . . . 'stays in' with a sub-
contractor." The system thus reinforced racial and skill divisions in Birmingham's coal industry, producing at the top a layer of subcontractors who made
"big money, without turning [their] own hand[s] to a lick of work"; in the
middle a layer of skilled men for whom "the ambition . . . to be a sub-contractor" precluded joint action with those below them; and at the bottom a
large mass of semiskilled and unskilled laborers, mostly black, who were
subject to "every sort of wrong, cruelty and injustice." Not being directly
employed by the operators themselves, laborers under the subcontract system had no access to the fruits of welfare capitalism. They were without protection against safety hazards, with operators able to "shift liability for accidents and personal injuries . . . to the shoulders of sub-contractors, many of
whom [were] irresponsible . . . or perhaps insolvent." Excluded from company housing, they lived "huddled together with their families" in "shoddy
groups of houses, shacks and huts, near the mines."[28]

Armes's account makes clear that the subcontract system was, like convict
lease, "distinctly and absolutely a product of anti-unionism" and not merely an arbitrary system of labor management. Prior to the 1904 and 1906

strikes, Armes recounted, it had been the "custom" in certain areas throughout the district for miners to employ "one laborer—or, if he had narrow work, two laborers, never any more." Beginning in 1904, however, in step with their protracted confrontation with the UMW, operators imported the subcontract system, "a West Virginia practice," in an explicit attempt to break the union. While the "best companies" in the district had developed "a more careful supervision" of the system by 1913, and certain operators had denounced it publicly as a scourge upon the district, many others—including prominent advocates of welfare work—continued to defend subcontracting as an essential component in keeping Alabama competitive.[29]

The freewheeling abuse generated under the convict lease and subcontracting systems inevitably fixed a low standard for the operators' treatment of black miners generally. Corporal punishment remained a routine element in labor relations throughout this period, and black miners bore the brunt of such mistreatment. They complained, almost universally, of the presence throughout the district of camp "shack rousters," whose job it seems to have been to drive free black miners to work on a daily basis. One Walker County mine veteran recalled that the operators had "fellers, company men that toted guns and [went] to see niggers . . . when they didn't come to work to see what the trouble was, and they'd tell 'em they'd better be there the next day." From Margaret mine came a protest that the company employed "what is called a company devil here and at all of their mines who rides around and sees that you do not lay off when you are sick." A black miner from Republic complained that "many of the camps" employed a "Shack Rouster" who was "almost inhuman in his treatment of the average Negro man and in many instances he takes unlimited authority with the Negro women." His words were echoed by a black miner in the pages of the *UMW Journal,* who complained that the "shack rouster . . . rides from sunrise until [sunset] with his billy and his revolver hanging to his saddle, and if any negro opens his mouth in protest of his cowardly acts with their women they are either beaten almost to death or shot down like a dog."[30]

The especially harsh treatment accorded to black miners was apparently not lost upon all whites in the coal camps. One white miner "of twenty-five years' experience" who had "worked in many mines and with many Negro miners" acknowledged that while "the white miner is treated bad enough" in Alabama mines, "the Negro as a general thing simply catches 'hell' in the big way." The operators "do not send 'Shack Rousters' to any white man's home to drive him out to work, sick or well, but day by day," he feared, "the reins are being drawn tighter and tighter on us . . . and if we continue to hold our peace the day will soon come when we will be catching as much hell as the Negro." The remedy, he concluded, was for white miners to "lay aside prejudice" and reorganize the UMW on an interracial basis.[31]

Lacking formal organization and deprived of legal remedies, black miners nevertheless attempted to set limits to the operators' transgressions. Given the formidable barriers to collective action, miners of both races most commonly registered their discontent through flight. Where possible, miners avoided those camps known for harsh treatment or moved along when they experienced abuse at the hands of a particularly abrasive mine boss or superintendent. Skilled miners, who could exercise some leverage, naturally gravitated toward those camps where a more moderate system of labor management prevailed. In rare circumstances, black miners attempted to stand their ground and challenge individual operators, but invariably during this period they ended up on the losing end in open confrontations with their employers. When, for example, a night boss at Sipsey killed a black miner in September 1914 for failing to show up for shiftwork, blacks there became "considerably wrought up over the incident." The superintendent, Milton Fies, reported that things "looked squally here for a while" until he imported a deputy from nearby Jasper and discharged "some five or six negroes, who were particularly impertinent and trouble seeking."[32]

Without any immediate prospects for redressing their grievances, however, black miners—as well as whites—took out many of their day-to-day frustrations upon themselves and each other. The squalor of the coal camps, the rootlessness of much of the mine workforce, and the atomization deliberately fostered by the operators all contributed to creating a coarse environment that reinforced the alienation of individual miners. "Alternate drudgery and dissipation," one observer wrote after a survey of negro camps in the district, "make [black miners] physical wrecks before middle age, so that that if it were not for the constant influx of new laborers from the cotton fields they would soon become extinct." A camp physician reported that the habits acquired by convicts remained with them after their sentences had been served. Having passed their leisure time in the stockades playing "dice and cards," "whole crowds" of black miners were frequently caught up in Sunday gaming sweeps. "Many of the women," according to this account, were "addicted to the intemperate use of strong drink," while cocaine was "frequently used by the more debased of each sex." Gunfire was a regular feature of life in the camps. Incidents like the one reported at Sayreton, where a black miner went on a "shooting frenzy" in October 1907, forcing "every house and store around the mines" to "[close their] doors," were typical. Even at TCI's model camp at Docena, the black communist miner Angelo Herndon reported in later years that "the men in their wretchedness grew brut-ish . . . [quarreling] with each other incessantly. The merest trifle was sufficient for them to leap at one another's throats."[33]

In such conditions, Herndon wrote, miners "did not crave for anything better. Hard work, maltreatment, ignorance and a bestial atmosphere con-trived to keep them blind as bats to their degradation. Every Saturday when they got their pay they went on a drunk and awoke . . . the following day more warped and embittered than ever." A black UMW veteran confirmed that camp life was rough. Blacks raised in the coal camps were commonly referred to as "camp nigger[s]," an epithet that marked them out as "rough jokers" who would "just . . . work and party and fight." Many operators considered the unsettled, footloose character of the mining camps an obstacle to stabil-ity. Anxious to attract skilled men, those employers who could spare the ex-pense devoted resources to ameliorating conditions in the camps. But in the long run, of course, they were attempting to recruit skilled men without adding considerably to their labor costs.[34]

The operators' determination to raise conditions in the camps while divert-ing only a minimal amount of capital to the venture cleared a path for the broad dissemination of the black middle class's doctrine of "racial uplift" throughout the district. Pieced together during those "dreary years" follow-ing Redemption, when "the advancing tide of segregation and disfranchise-ment made protest seem futile," this strategy rejected the possibility of direct confrontation with Jim Crow and looked instead to individual moral reform as the route out of poverty and oppression. Shaped by the twin imperatives of "Jim Crow terror and New South economic development," Kevin Gaines has written, uplift "transformed the race's collective historical struggles against the slave system and the planter class into a self-appointed personal duty to reform the character and manage the behavior of blacks themselves." As it came to be applied in the coalfields, middle-class accommodation prescribed as a solution to the problems of black poverty and powerlessness individual thrift, faithful service to one's employer, sobriety, and self-discipline.[35]

Such a message was obviously compatible with the maintenance of a low-wage, nonunion regime in the coalfields, and well before 1908 district oper-ators had undertaken to develop close ties with Birmingham's small but in-fluential black middle class. H. C. Smith, a conservative black Democrat, told interviewers that he had made a living from just after the turn of the centu-ry by performing "welfare work" under the auspices of the "mining inter-ests of Birmingham and vicinity." The founder of an organization calling itself the Southern Afro-American Industrial Brotherhood, Smith verified that coal operators had supported him sufficiently in his efforts to "fight the

organization of the Negro workers" into the UMW to relieve him from having to "worry for [financial] support." During the 1908 strike operators had
welcomed a steady barrage of antiunionism from the pages of the Reverend
William T. McGill's weekly *Hot Shots,* a black-owned newspaper widely distributed in the mines. "Colored workmen" who worked through the strike
and "remained loyal to the company" were lauded in the paper while those
who struck were denounced as "good-for-nothing indolent vagabond[s]"
who by their disloyalty had brought "disgrace [upon] the race." "Why abuse
the companies' good graces?" McGill inquired of his readership as the district trembled on the knife-edge of racial conflagration. "The negro should
wake up, quit his foolishness and go to work."[36]

Coinciding with their reorientation toward native mine labor and their
embrace of welfare capitalism after 1908, district operators hastened to shore
up their ties with Birmingham's black middle class and to involve it at the
center of the paternalist project. In this effort they enjoyed a unique advantage in the unmatched hegemony of Booker T. Washington's accommodationist outlook among the district's most prominent "race leaders," a perspective that dovetailed perfectly with the role capital had earmarked for
black workers. Washington never tired of reminding his patrons among the
region's industrial elites that black labor alone guaranteed them immunity
from the "strikes, lockouts, and labor wars" common in the North. Rather
than shun the strikebreaking role that capital seemed intent to impose upon
black workers, Washington embraced it as a positive advance, calculating that
their willingness to underbid white labor would win for black workers a
permanent foothold in industry. Inevitably, the accommodationists rejected trade unionism as a vehicle for black advance.[37]

Washington did not pioneer industrial accommodation; instead he popularized sentiments that had been taking shape among the South's black
middle class since the collapse of Reconstruction. Nevertheless, his personal
prestige played an important role in heralding the gospel of industrial accommodation through the Birmingham district. He had been a frequent visitor
to the city and wielded considerable influence among Birmingham's black
establishment. A few prominent race leaders had attended the Tuskegee Institute, while others were active in his National Negro Business League. One
of them, William Pettiford, spoke frequently at its annual meetings and had
been endorsed by Washington for the position of assistant commissioner for
the Atlanta Exposition in 1895 and later as U.S. revenue collector for Alabama.
He was eventually hired to coordinate welfare work among black miners for
TCI. In the early years of the twentieth century, Washington had even col-
laborated with local black elites—including Pettiford—in an attempt to

launch a black-owned coal mining company in the district (apparently with assistance from TCI). Affirming Washington's authority in the district, a white city official recalled approvingly that "local leaders with like ideals" were "numerous," with the result that the "Birmingham negro, guided by [such] wise leaders, has found his groove and quietly moves within it." NAACP officials who were sent to the South to organize branches in 1914 found Alabama "utterly Booker-T-Washingtonized . . . and therefore not kindly disposed toward us," and while they succeeded after some effort in building branches at Selma and Montgomery, none had been formed in Birmingham as late as 1918.[38]

Birmingham during this period was not merely a stronghold of accommodation: its unique constellation of powerful industrial interests, an articulate black middle class, and the largest concentration of black industrial workers in the nation made for the most systematic application of Washington's labor relations formula anywhere in the United States. Hampton and Tuskegee graduates played a prominent role in the various "negro welfare associations" set up in the camps, and operators hired them to teach in and administer their Negro schools. Sipsey management appointed the Tuskegee protégé Robert W. Taylor to oversee the education of black miners' children, considering him a "good negro, smart," who "knows his place." W. B. Driver, the chairman of the Jefferson County Republican Club, and Oscar Adams, the editor of Alabama's largest black weekly, maintained an ongoing relationship with leading operators, frequently sharing the podium at management-sponsored meetings with company officials. Mine owners closely vetted ministers in the company-sponsored camp churches to assure their soundness on the "labor question." The practical value of such relations in deflecting black workers' grievances was illuminated in the memoirs of the TCI manager James Bowron. Meeting regularly with black community and religious leaders in the offices of Birmingham's American Bank to "discuss the status of the colored race amongst us and to consider what the difficulties were," Bowron recalled "advocat[ing] their being taught thrift by their preachers and avoidance of gambling, so that they might save some money to put into their houses."[39]

Spurning trade unionism as a vehicle for racial advance, Washington's followers beseeched black workers to place their hopes in sound relations with the "better class of white men" and loyalty to those who paid their wages. "It should be the motto of every colored laborer," a typical editorial suggested, "to strive to make friends with his employer. . . . Alabama has the richest soil, the friendliest skies and the best opportunity for advancement to the negro that can be found anywhere." Whether a black laborer prospered or failed

depended "not on how much he will be recognized in a union," another reasoned, "but how well he can please the men he is working for." In the view
of Birmingham's most prominent race leaders, strikebreaking was not only
a legitimate activity for blacks but presented the best prospect that black labor might secure a monopoly on manual labor in the district. "Give the Negro
the laboring field, give him the contracts in the mines, steel plants and many
other fields," Adams proposed to industrial employers, in a tone reminiscent
of Washington's, "and strikes will cease."[40]

It is tempting to view the close relationship between the district's industrial employers and middle-class "race leaders" as being driven by mercenary
interests or sheer opportunism, and there is ample evidence that both factors played a substantial role in individual cases.[41] But the wider context of
ubiquitous white hostility on the one hand and the distinct petit bourgeois
orientation of the black middle class on the other hand was far more fundamental to the process. The black middle class's rejection of interracial unionism as a solution to the plight of black workers was in part a pragmatic
response to the weakness of such an alternative in the South and the less-
than-exemplary record of the national labor movement in relation to black
workers. Certainly the craft-based steel unions in the Birmingham district,
which hinged their success (of which they did not enjoy much, it should be
noted) on a policy of racial exclusion, merited the disdain not only of uplifters but of the black working class generally.[42] The black middle class's
hostility to the UMW in northern Alabama, however, where employer opposition invariably centered on the union's commitment to racial egalitarianism, suggests that there were other fundamental pressures shaping their
outlook.

None of the district's most prominent race leaders—men like Oscar Adams, W. B. Driver, the Alabama Penny Savings Bank founder William R. Pettiford, the pastor, building and loan association founder, and occasional coal
mine operator T. W. Walker—were descended from the region's small free
black community. Most had been born slaves or the sons of slaves, and the
majority of them had accumulated whatever wealth they owned by offering
services to the black community that whites refused to provide. The pattern
noted by August Meier of a "growing antipathy on the part of whites toward
trading with Negro businessmen" leading to a "real burgeoning of Negro
enterprises after 1890 . . . based chiefly on the Negro market" is confirmed by
developments in Birmingham. The barriers erected by segregation had cre-
ated a distinct black business enclave, with a single building in the "Negro
quarter," the six-story Knights of Pythias building, housing within its walls
"most of [the city's black] professional[s], doctors, dentists, and lawyers."[43]

By its nature, white supremacy flattened class distinctions on both sides of the color line, reducing blacks from all levels into a single amorphous mass in the eyes of white southerners. In doing so it obscured the real material, cultural, and social differences that divided black Birmingham. While the city's black middle class was not immune from the range of indignities attendant with racism,[44] their wealth and standing insulated black elites from its most unsavory effects. "The main targets" of racial violence were "poor black men, the foundation of the southern working class," Nell Painter insists, so much so that "the victimization of prosperous black men . . . was almost incidental to the immobilization of millions of black workers." Dealing almost exclusively with either black clientele or wealthy white patrons, black elites seldom rubbed shoulders with poor whites, as black workers did on a daily basis. Like ordinary blacks, they expressed no burning desire to mix socially with whites, particularly when such relations were saturated in humiliation.[45]

Unlike black industrial workers, however, for whom segregation usually meant confinement to the least remunerative, most dangerous occupations in the mines and mills, the black middle class was in significant ways materially dependent upon the maintenance of the color bar. A central element of their prescription for uplift was race patronage of black businesses. "Patronize your race institutions first, last, and all the time," F. P. McAlpine, the editor of *Free Speech,* exhorted his readers in 1903. The "Buy Black" slogan was a constant theme of uplift ideology throughout the early twentieth century, widely disseminated through the black press and through Washington's National Negro Business League. The black entrepreneur R. S. Burlong's 1915 address to that body, in which he lamented the tendency of black workers to spend their wages outside the "Negro quarter," was typical: "Every Negro must be made to realize that every dollar invested in Negro enterprises stimulates the life of these enterprises, and as a result . . . Negro business is made more self-sustaining." W. R. Pettiford believed likewise that the "colored wage-earner" had to be "prevailed upon to spend his earnings so that a portion . . . may be retained by his own people." Like many of their class, these men could not understand that the disloyalty they detected among poor blacks was a pragmatic response to the ability of larger, white-owned enterprises to undersell them.[46]

At its most benign, the strategic transformation in race thinking during the years following Redemption expressed itself simply as an embrace of the regnant social Darwinism trumpeted by defenders of the new industrial capitalist order everywhere. On one level, there was nothing remarkable about the fact that the black middle class shared the faith of their white counterparts in the ameliorative powers of the market and the goodwill of leading industrialists. But there was something bizarre about their easy separation of economic dogma from the frequently brutal reality engulfing the great mass of black southerners. In its unadulterated form the doctrine articulated by the *Birmingham Reporter* editor Oscar Adams, that American society was a "great highway" upon which "it isn't so much color, previous condition or race, [but] who you are" that determines "whether you will . . . be respected," could never have won more than a tepid reception among black workers; nor could McGill's exhortation to black miners during the course of the 1908 strike to stop their "[constant] grumbling about the white people not paying us for what we do" or his insistence in an editorial three years later that "at none of the [steel and iron] plants is the colored man discriminated against in any way or manner."[47]

The compulsory nature of many of the activities that brought "race leaders" into contact with rank-and-file black miners, invariably directed by company officials, makes it difficult to gauge the actual resonance of uplift. To the extent that Adams and others succeeded in winning a following in the camps, however, it was their espousal of a form of black racial pride rather than the appeal of middle-class morality that earned it for them. In the absence of an interracial alternative even the cramped variant of racial pride advanced by uplifters could strike a chord among hard-pressed black miners.[48]

Aside from the company-sponsored welfare associations, a number of other, semiautonomous race institutions played a key role in cementing the relationship between race leaders and the objects of their efforts. Of these, the most important were the numerous black fraternal orders, which, though usually directed from the outside by notable "race leaders," won a wide following among men and women in the camps.[49] Against a backdrop of formidable impediments to black civic participation under Jim Crow, blacks throughout the South looked to fraternal orders like the Colored Odd Fellows and the Colored Knights of Pythias for "unhampered opportunity for social life and . . . the exercise of leadership." By their seeming innocuousness, the orders provided a vehicle for race association acceptable to whites in a society obsessed with fears of race insurrection. The fact that they also offered sickness and death benefits to members enhanced their appeal among miners who otherwise enjoyed little financial security. Above all, however, the popularity of the fraternal orders in the camps reflected the appeal of racial pride: "Here only," a mining camp physician observed, "has the negro his best opportunity for self-government. There is no excuse for hiding the sins of a brother from fear of too harsh treatment at the hands of another race." The élan exhibited by fraternity members can be gauged from the fact that "some white men regard[ed] them with distrust, thinking that this banding together is for offensive rather than defensive purposes." No doubt many white Birminghamians shared the paranoia expressed in a report from neighboring Georgia, where officials regarded the fraternal orders as "nothing more nor less than hotbeds of anarchy and Bolshevism . . . always plotting and scheming against the white race."[50]

While white hostility lent itself to the internal cohesion of race organizations, the black middle class's "missionary notions of uplift" and their close ties to the coal operators could not help but generate intraracial division. The black middle class's petit bourgeois vision, which aimed to remold the character of individual blacks rather than challenge inequality, suffused the work of the fraternal orders, the black church, and the camp welfare associations. Their elitism bordered at times on outright contempt for the less fortunate of the race. The most important function of the fraternal orders, from the perspective of black elites, was not their dissemination of racial pride but their role in "counseling against idleness, extravagance, and indifference." Race leaders more often stood closer in social terms to white elites than they did to black workers. The more successful of them worried openly that the black working class and the poor would "drag [them] down." Adams's *Reporter* warned his readers incessantly to stay clear of the "unworthy Negro," the "Negro swell," the "Negro gambler . . . of the crap game gentry," and the "dishonest Negro," all of whom seemed to be "conspiring to pull down what the 'worthy' of the race had built up." The dissemination of middle-class morality figured prominently, therefore, in their intervention in camp life.[51]

The most troublesome feature of camp life, from the perspective of some black elites, was the seeming laxity of sexual mores.[52] "Marital ties are regarded very lightly in many instances," one observer noted. "I have seen men living with as many as three different women in as many years. . . . Some marry each new wife without divorce, often under an assumed name, while others are never married to any of the women." The black miner Bobby Clayton recalled that the tendency of miners to move about in search of work rendered family life unstable: "some men had two wives and this one stayed . . . here and one stayed somewhere else and . . . the women would get along together. Men had double families and wasn't no conflict." In a situation where "the churches wield[ed] only a limited influence," the female divisions of the fraternal orders became noted for their efforts to rescue black women from sexual immorality. The matron of Birmingham's black establishment, Adams's mother-in-law Carrie L. Tuggle (eulogized in the local white press as the "female Booker T. Washington"), called for a "higher and nobler womanhood among the Negroes" and urged black women, whom she apparently believed to be engaged in prostitution, to "spurn fine dress not obtained in an honorable way."[53]

These efforts involved an important element of racial self-preservation. The Court of Calanthe, the female auxiliary to the Colored Knights of Pythias, endeavored especially to "[prevent] the further amalgamation of the races thru the young negro women." Such efforts, typically adopted as defensive measures against the subjection of black women to unwelcome advances from white men, were frequently endorsed by black miners, including union supporters. Nor were miners of either race hostile to imposing moral order on the camps. A Republic miner who believed that "the Negro miner[s] almost to the man recognize . . . that they should be organized" complained in the same letter of "men and women being allowed openly to live together without being married . . . in lots of cases run[ning] blind tigers." He worried that it was "almost impossible" to raise children "to be useful men and women" in such an environment and looked to the union (rather than the employers or the black middle class) for a remedy to this situation.[54]

Still, class tensions seem to have been just as capable of intruding upon relations between female uplifters and the objects of their reform work as they were among men. Lynne Feldman found in her pioneering study of black Birmingham a "rather sharp delineation of fraternal membership among [black] women from different socioeconomic classes." The Court of Calanthe "practiced an exclusionary membership which kept the order free from a lower class of women" and, like other middle-class black women, "removed themselves from more democratic organizations." One telling sign of the potential fracture lines that such a policy could result in occurred as a result of the elite voting rights strategy pursued by race leaders. In 1920, eight hundred to a thousand black women were given the franchise after W. B. Driver "with a few friends interceded for them . . . with certain prominent members of both races," an effort in which he reportedly enjoyed "the thorough cooperation of the best white citizens of the city." Tensions erupted, however, when other black citizens, described by Adams as "agitators and race disrupters," "hindered" the registration. The delaying of the registration he attributed to "the same old worn-out-do-nothing crowd . . . antagonistic to the progressive spirit." In truth, Adams was far from an impartial observer. His mother-in-law, Carrie Tuggle, was the first black woman to register to vote in 1920 and one of two prominent African Americans assigned to vet black applicants, chosen for the position because they were "deemed respectable by whites in power." In the end, a mere 454 blacks were granted suffrage in Jefferson County in 1920, less than a quarter of whom were women.[55]

Race organizations could only balance the contradictory interests of black miners and race leaders in periods of relative calm and when interracial alternatives were too weak to disrupt the racial cohesion encouraged by black elites. But tensions did erupt within race organizations, almost uniformly around the central problem that uplifters refused to confront: the exploitation of black workers at the hands of their employers. Washington's National Negro Business League, described by one historian as "the organizational center of black conservatism," had conceived of itself as a "clearinghouse for the employment of all kinds of skilled and unskilled black labor," and its local supporters came eventually to share the frustrations of area industrialists at the unreliability of black workers. Adams recounted his exasperation at attempting to find dependable workers to fill openings at a plant employing black labor exclusively. "We sent twenty-five or thirty men to the factory," he recalled, but "not a third of them remained, not half of them began to work." Unable to contemplate any other reasons why workers might walk away from such a generous "opportunity," Adams attributed the disastrous outcome to their failure to "see the need for so much money at the loss of their usual frolic."[56]

Despite their being led from outside the coal camps by representatives of the black middle class, in every major crisis black miners transformed the fraternal orders into vehicles for spreading the gospel of trade unionism, a development that threatened to tear the orders apart. During the 1908 strike, Henry Claxton Binford, the grand master of the Colored Masons, warned UMW supporters in his ranks that he "consider[ed] every Mason in the Birmingham district connected with these unions . . . a murderer" and vowed that he would deny death benefits to "any Mason killed while affiliating to these unions." Binford's diatribe provoked a sharp response from a black UMW member, who replied that he "cannot swallow every camel that Binford and the SCM tries to force down my throat. He is not the whole cheese so far as colored Masons are concerned." The miner's angry rebuke ended with the telling assertion that "the miners' union has done more for us than all the secret orders combined, and Binford thrown in."[57]

Black miners' conditional embrace of "uplift" therefore reflected the appeal and limitations of a doctrine that championed the racial integrity of African Americans but emphatically rejected any attempt to link their predicament in the camps to exploitation by the operators. To the extent that "race leaders" like Binford had identified themselves with the interests of the mine owners, such tensions were inherent in relationships with their constituency in the camps. The fault lines could be seen, for example, in the tensions between black miners and the churches. Many ministers active in welfare work, particularly in the black community, were financially dependent upon the operators, and their sermons seldom deviated from the innocuous themes palatable to management.[58] One black miner complained, for example, that many preachers were "nothing more than stool-pigeons for the coal companies" who, "instead of preaching the Gospel of the Son of God," espoused "the doctrine of union hatred." As a result, churches in the mining communities often drew the bulk of their congregations from "among school teachers, company officials, and adjacent farmers" rather than working miners, who "preferred not to worship with mine officials or office men in the same congregation." The coal operators, another miner reckoned, "would rather pay fat salaries to these ministers than to recognize that labor has any rights to a decent pay for a day's work."[59]

Uplifters were clearly disturbed by the lukewarm support accorded the established churches in the camps. The proliferation and popularity of itinerant preachers, many of them miners themselves, led P. Colfax Rameau, the editor of the black-run, ardently antiunion *Workmen's Chronicle,* to call upon the operators to drive them out of the camps. "Why should there be in one camp where there are not more than two or three hundred communicants two and sometimes as many as three or four Missionary Baptist Churches?" he inquired. Rameau accused the "bunch of Negro ministers that are running down to the different camps" of being "in the Ministry for the 'Fishes and Loaves.'" Frequently, prominent uplifters reserved their harshest invective for those churches committed to an emotional service. "The minister can no longer lead his people with spasmodic sermons and playing upon the feelings of his hearers," a company-supported pastor from Garnsey complained. "He must be prepared to lead his people intelligently."[60]

The pervasiveness of racism went a long way toward blurring class distinctions in the black community. The racial antipathy harbored by white miners, latent throughout the South but aggravated to some extent by the operators' newfound "preference" for black labor, encouraged black dependence upon mine owners and reinforced racial paternalism. The absence of a viable interracial alternative precluded collective protest, even where black miners felt their grievances keenly. But the disparity between the operators' claims and the actual conditions under which black miners lived and worked meant that sooner or later the system would give. Black miners grew increasingly impatient with the accommodators' motto, "Festina Lente" (hasten slowly). They grew "tired of the beautiful promises made us by the operators" and complained openly that their "leaders" were "selling us out to the bosses for a mere mess of porrage [sic]."[61]

The fundamental weakness of racial paternalism was that it never delivered on its promise to those seeking refuge from the poverty and despotism of Black Belt agriculture. There was work to be had—plenty of it—in Birmingham district coal mines, but only under terms that kept the bulk of black miners and their families living in poverty and under a regime that reproduced the racism and brutality so familiar to those fleeing the Cotton Belt. The U.S. Immigration Commission found in 1911 that "as a general rule, negro miners occupied a poorer kind of house [in the camps] than either the native white or the immigrant." Segregation remained "very pronounced," and black miners continued to be paid substantially less than whites for comparable work. No system bearing such serious flaws could succeed in holding the permanent allegiance of black miners, and by early 1916 the disaffection of blacks from the paternalist system was becoming obvious. They only awaited a breach in the operators' control to assert their rights.[62]

4. White Supremacy and Working-Class Interracialism

THE BENEVOLENCE THAT DISTRICT OPERATORS extended to their black employees cost them little and promised them much. Above all, the special relationship that they cultivated with Birmingham's black middle class—and through them black laborers—represented a long-term investment in labor peace. With the UMW hounded out of the district, the mine workforce split along racial lines, and wages reduced to a level that could only be imposed on the dispirited and disorganized, operators could afford a degree of magnanimity in their dealings with black workers. The paternalist tradition provided a framework that allowed mine owners to represent themselves simultaneously as "friends of the negro" and as staunch defenders of white supremacy. But while leading white employers throughout the South "had little to fear from the negro and . . . considerable to gain by establishing themselves in a paternalistic relationship as his protector and champion against the upland and lower-class whites," the white rabble that is presumed to have filled the ranks of the Ku Klux Klan during Reconstruction or the lynch mobs during the years straddling the turn of the century[1] frequently found themselves locked into desperate competition with blacks. This was certainly true for white miners in the Birmingham district during the years after 1908.

The bitter, intense rivalry at the bottom of southern society was frequently encouraged by those at the top. "In nearly all the trades," one historian of the New South wrote just after the turn of the century, "the rates of compensation for the whites is governed by the rates at which blacks can be hired." Pay rates for unskilled southern whites were "almost as low as the wage for blacks," another noted, and wage levels in the textile industry and in mining and manufacturing centers were closely tied to black agricultural wages.

It was not merely the proximity or abundance of cheap black labor that planters and New South industrialists took advantage of. Their ability to impose low living standards upon working southerners of both races was greatly enhanced by their proclivity, in the event of a strike, to "hold the great mass of negro mechanics *in terrorem* over the heads of the white." In this situation, racial animosity was built into the very structure of the industrial South. More than fifty hate strikes launched by white workers against the employment of black southerners between 1882 and 1900 demonstrated not only the firm hold of racism upon whites locked out of New South prosperity but also the failure of Jim Crow to assure white working people of "an easy superiority over their black counterparts."[2]

The white South's seemingly unanimous resolve to uphold white supremacy rested upon a number of volatile contradictions. In a region where the immense disparities between poverty and wealth were transparent and where "the two races were brought into rivalry for subsistence wages in the cotton fields, mines, and wharves," C. Vann Woodward has written, "it took a lot of ritual and Jim Crow to bolster the creed of white supremacy in the bosom of a white man working for a black man's wages." Moreover, white supremacy did not—and could not—mean the same thing to poor whites as it did to white planters or industrial employers. The ascendant racialist ideology around which disparate (and potentially antagonistic) elements of the post-Reconstruction white South cohered was "not simply a summary of color prejudices," as Barbara Fields has written, but contained "a set of political programs, differing according to the social position of their proponents." The racism of the "rich black-belt Democrats was annexed to an elitist political ideology that challenged . . . the political competence of the subordinate classes as a whole," while that of the poor whites lent itself to "political ideologies hostile to the elitist pretensions of the black-belt nabobs and, at least potentially, solvent of some of the grosser illusions of racialism."[3]

The record of organized labor in the South is replete with evidence that antiblack racism suffused the general outlook of southern white workers. Before the rise of the Congress of Industrial Organizations (CIO), such labor movement as existed in the South was dominated by the narrow, bread-and-butter unionism typical of skilled workers' organizations elsewhere throughout the United States. Craft unionism, embodied in the cautious strategy pursued by the American Federation of Labor, operated upon the logic that the unions could best position themselves to influence wages and working conditions through tight regulation of the skilled labor market. Rather than undertake a broad campaign to organize all workers—skilled and unskilled, across craft lines—in a particular industry, craft organizations

directed their efforts at select, highly skilled groups of workers thought indispensable to production. At their best the craft unions might succeed in
organizing a number of autonomous union locals representing the various
skilled trades in a given plant, with the bulk of unskilled workers completely excluded from representation. While local AFL affiliates experimented with
a variety of independent political formations and were on occasion pushed
into organizing industrial workers, in general they fought an uphill battle
against a leadership that threw its weight against every manifestation of independent working-class politics, accepted the basic framework of American capitalism, and believed that economic action alone would secure for
workers their just share of industrial prosperity.[4]

In the North and the West, the exclusionary policy that flowed from craft
unionism weighed heaviest upon working women and on the growing population of new immigrants, who were employed disproportionately in industrial settings at semiskilled and unskilled positions. In the simmering racial
cauldron of the South, however, exclusion fused easily with Jim Crow to produce a strain of "lily-white" unionism unmatched elsewhere in the United
States. Political abstention left the Democratic Party's hold over the loyalty
of southern white workers unchallenged, and with few exceptions black
workers remained alienated (and often constitutionally barred) from the
official labor movement. Although a small minority of white trade union
militants—mostly individuals associated with the syndicalist Industrial
Workers of the World (IWW) or with the Socialist Party—upheld a princi-
pled opposition to racism in the unions, white supremacy was so ubiquitous
that it infected the left wing of the labor movement as well. Southern members of the Socialist Party, for example, refused to broach the thorny issue of
"social equality" and saw no contradiction between their ostensible commitment to working-class solidarity and their assumption—even insistence—
that racial segregation would be maintained under a new, worker-led socialist
government. To the extent that the fragile shoots of working-class organization extant in the South during this period can be said to have constituted a
labor movement, their practice and outlook accommodated more often than
it challenged the region's embrace of white supremacy.[5]

The economic constraints under which the industrial South came of age
placed a premium on the employers' ability to manipulate racial divisions,
however, periodically upsetting the notion of white caste solidarity and compelling white workers to rethink their relationship with black labor. In a situation where employers harbored few qualms about displacing white labor with
black when such a policy would reduce labor costs, "two possible but contradictory policies" presented themselves to the established labor movement.

According to Woodward, white workers were forced to either "eliminate the Negro as a competitor by excluding him . . . or take him in as an organized worker committed to the defense of a common standard of wages."[6]

The exclusionary policy dominated southern white labor's defensive strategy throughout the late nineteenth and early twentieth centuries. By 1901, a correspondent to the *Bricklayer and Mason* related that blacks' standing in the skilled trades had actually declined with the advance of trade unionism. "Before organization," C. L. Berry wrote, "colored and white mechanics have been known to work side by side," but "when the union was formed, the colored men were rigorously excluded from membership." Some unions formally inscribed race discrimination in their bylaws. The Railroad Brotherhoods, for example, played a despicable role in urging southern employers to remove blacks from skilled positions and included in their constitutions provisions that limited membership to employees "born of white parents, who [are] sober, moral and otherwise industrious." This was apparently not enough to satisfy members in Amarillo, Texas, however, who submitted for approval a constitution that included the provision that "No member shall work on any job with a negro or Italian."[7]

The failure of exclusion to stall the aggressive campaign being waged by southern employers forced some in the labor movement to reexamine their resistance to organizing black workers. After investigating an appeal by white trade unionists from Augusta, Georgia, that blacks be excluded from their ranks, M. P. Carrick of the Painters' Union concluded not only that the demand should be resisted but that "the negro was the best union man." "The white people of the South will have to understand that if they ever want to improve their conditions, they must extend the hand of fellowship to the negro worker," Carrick wrote to the AFL president Samuel Gompers. "He enters into competition with them, and the bosses do not care who works for them, so long as those who do work . . . will work cheap."[8]

The gravity of southern labor's "race problem" can be gleaned from the fact that the debate found its way even into those unions that had long stood as bastions of conservatism and racial exclusion. The Carpenters' Union, whose Birmingham local had earlier spurned an attempt by blacks to affiliate (even denying them "any assistance in organizing a Local of their own"), in 1903 appointed a black organizer for the South, a decision that provoked wide-ranging discussion in the union's journal. A number of writers rejected any attempt to tamper with labor's color bar and held to a policy of exclusion. Others recognized, even if reluctantly, that so long as blacks remained outside the ranks of organized labor southern employers would retain the ability to deploy them against attempts on the part of white workers to raise their own living standards and therefore called upon the union to open its doors. The Negro was "preferred" by contractors, one correspondent explained, "because he can do, or be driven to do, a larger days' work, and for less pay. He will work longer hours and for less pay without objecting, and on his wages and hours the standard is set for the white mechanic." The only way to remedy this, he concluded, was to "lay down all our prejudice [of which, he admitted, 'I have as much as any southern born man'] and look the question fair and square in the face. . . . The negro is a friend to us on the inside, while on the outside he will be our enemy." Another union carpenter responded in a similar vein, insisting that the union open its doors to blacks and brushing aside the charge that such a policy would represent an affront to southern racial custom. "If we can afford to work all day on a scaffold beside them," he wrote, "then we can surely afford to meet them in the hall for an hour or so once in a while."[9]

Pragmatic interracialism of this variety did not compel white trade unionists to reject the accepted tenets of white supremacy. Though generally distancing themselves from the coarse, overt racism manifested by outright exclusionists, advocates of pragmatic interracialism almost invariably based their appeals for biracial cooperation in terms of white workers' self-interest. Black workers required organization, it seemed, not so much for their own good as for the effect that such organization would have in safeguarding white workers' living standards. Though occasionally lacing their appeals with sympathy for the exploited black laborer, white trade unionists typically adopted a tone that was more patronizing than egalitarian. The carpenter quoted above attributed blacks' willingness to "stand more abuse" to their lack of "the finer qualities of nature and self-respect" and their "not considering [themselves] on an equal footing with the white man." Thus it became incumbent upon the ostensibly more advanced white workers to "take the negro by the hand" and "raise" him to the standards of the white.[10]

In the highly charged atmosphere of the Jim Crow South, however, even the limited interracialism advanced by a minority of skilled white workers was sufficient to provoke the animosity of employers and their allies in the upper ranks of white society, and their desertion by respectable whites led erstwhile pragmatists to bold if unsustained defiance of racial custom. When hotel owners refused to rent space to Alabama's interracial UMW for its 1901 convention, union officials ordered a boycott of the facilities until owners acquiesced and apologized. A year later, when the Alabama State Federation of Labor attempted to hold its convention at Selma, the local press objected to the fact that its black vice president, J. H. Beane, had been trusted with overseeing arrangements for the gathering and baited its readership with the specter of an integrated meeting taking place in their community. White delegates refused to compromise on the issue, however. "Rather than see one accredited delegate, black or white, 'thrown out' of this convention," a member of the (otherwise notoriously lily-white) Birmingham Typographical Union declared, "I would go to the woods and hold this meeting." In the end, the convention was allowed to proceed after a local chapter of the United Confederate Veterans offered use of their hall "without regard to the color line."[11]

The vigilance with which white supremacy was enforced infringed upon even the limited interracialism envisioned by pragmatists, and in the course of defending elementary economic interests they were frequently led—in spite of their own deeply held prejudices and their formal commitment to white supremacy—to defy the southern caste system. For a small minority of white workers the transparency of the employers' racial strategy and the barriers erected against defense of their basic economic interests led them to question the very foundations of southern society. For some attracted to a left-wing critique of southern racism, their experience led them to reject white supremacy as a crude hoax being perpetrated against the poor of both races in the interests of capital. A Birmingham Socialist Party member urged readers of the *Labor Advocate* to "refuse to be divided" by those attempting to "keep themselves in power by making the common herd fight among themselves." "Joy prevails around the table of the National Association of Manufacturers as long as the Jew is cursed by the gentile," he wrote, "and the railroad magnate is blissfully content as long as the white man considers the negro as his industrial enemy."[12]

Occasionally the radical critique extended to an indictment of the otherwise sacrosanct Democratic Party, as it had after Governor Comer's smashing of the 1908 miners' strike. "Will you kindly point out to me where the democratic party in its attitude toward labor has ever been anything but a party of hot air, frothing and deceit?" an outraged correspondent chided *Advocate* subscribers. "Don't you know in this Southland of ours where the Democrats have ruled supreme for forty years there is not a single law on the statute books for the protection of the workers—that we are completely ignored like so many cattle?" "The day has passed when we men of the South can be fooled by the hackneyed cry of foreigner, carpet-bagger, alien, assassin, and agitator," wrote another, somewhat prematurely. "Our fathers were foreigners and we are carpet-baggers."[13]

The caste system so deeply entrenched in the early twentieth-century South was deeply rooted in the history of the region. What labor movement existed was confined to a handful of major industrial towns and cities and never enjoyed a period of relief from the consuming struggle to maintain a basic existence. And within the ranks of organized labor the racial ideology regnant throughout every strata of the white South resonated among white workers willing to blame their precarious social standing upon those even less fortunate than themselves. Where considerable numbers of whites were willing to depart at all from the policy of racial exclusion, more often than not they did so without breaking in a fundamental way from white supremacy. The small minority of union militants willing to go a step further were greatly outnumbered, pilloried by representatives of the "better class of whites" and often by their fellow unionists as traitors to their Anglo-Saxon heritage.

Nevertheless, the acute social divisions in white society continually undermined the foundations of white racial solidarity. The impulse toward interracial cooperation, even when it succeeded in surmounting the prejudices of white workers, was frequently squashed or contorted under the weight of broader white hostility. Yet it remained a permanent if underutilized option in the southern working class's search for an effective defense against the New South industrial juggernaut. While before the mid-1930s interracialism remained a rare phenomenon, at certain junctures a complex but vibrant interracial tradition took root, posing a sharp alternative to the dominant lily-white tradition and eliciting a frantic response from the guardians of white supremacy. On the New Orleans waterfront, in the logging camps and sawmills of western Louisiana and East Texas, and in the coal and iron mines of the Birmingham district black and white workers built organizations that defied southern racial protocol and challenged the prerogatives being fastened on the region by leading men of the New South.

Well before the founding of the Birmingham district, the white yeomen farmers inhabiting the sparsely settled foothills of northern Alabama had developed a reputation for defiance of the powerful Black Belt planters who controlled state politics. Opposition to the so-called nigger lords did not necessarily translate into sympathy for their black chattels, however. Like nonslaveholding whites throughout the South, Alabama's yeomen "nourished a deep antipathy almost equally directed against the institutions of slavery, the planters, and the blacks, all as symbols of an unfair economic competition they found it difficult, if not impossible, to sustain." During the crisis that erupted eventually in the Civil War, all seventeen counties in northern Alabama sent cooperationist delegates to the secessionist convention, a majority of whom voted against secession. And by the war's end, the region had won notoriety among southern nationalists as a hotbed of unionism and "a well-known refuge for Confederate deserters and stragglers, who came and went without fear of arrest," while "Confederate army enrolling officers could expect to be ambushed and killed."[14]

The antagonism between white yeomen and elites carried over into the period following Reconstruction, exacerbated by the disruption of rural life under the impact of industrialization and the intrusion of the market. The transformation of the area that would later comprise the mineral district added two important elements to the region's tradition of class dissent. The much-heralded entry of the "Big Mule" industrialists into state politics and their alliance with the yeomen's traditional Black Belt antagonists reinforced the alienation of hill-county whites. No longer were they dominated by a distant foe, whose aggrandizement at their expense was taking place in legislative halls at Montgomery. Rather, in the mines sprouting up around the emerging industrial district and along the railroad tracks that carried the output of these operations to markets or mills in Birmingham and further afield, native whites confronted directly the muscular exercise of economic power that they themselves so obviously lacked. Not surprisingly, the region became host to a series of formal challenges to elite control, directed now against the Big Mule–Black Belt alliance wielding power at Montgomery.

The character of these insurgencies was indelibly shaped by another by-product of industrialization: the migration of large numbers of African Americans into the coalfields in search of work. Alabama blacks had populated the Tennessee Valley region to the north of the mineral district since well before the war, but it was the mine owners' incessant demand for labor that initiated a steady influx into the coal-producing counties after Redemption in 1874. Growing increasingly disenchanted with their treatment in the Republican Party (which was fast shedding its commitment to black civil rights and attempting to recast itself as a lily-white organization), blacks working around the mines found much to support in the appeals of the independent movements springing up in the district. At their best, white hill-county anti-Redeemers were astute enough to understand that black electoral support could not be secured without a genuine attempt to address the concerns of black voters. There were of course many issues upon which poor whites and blacks could close ranks without any special effort. "As increasing numbers of hill-county farmers suffered from the political and economic developments of the post-Reconstruction era," Michael Hyman notes in his important study, "[class] issues contained the potential for bringing black and white ordinary people together to deal with matters of concern to both." Though "primarily concerned about the rights and liberties of whites," he contends, the independents' "opposition to labor and market regulatory measures aided black as well as white southerners."[15]

More impressive even than the potential for class issues to submerge the region's fixation with race was the tendency of committed white dissenters to go beyond straightforward economic questions and shoulder a commitment to the broad defense of black civil rights. Their involvement in interracial struggles over bread-and-butter issues led significant numbers of white anti-Redeemers "at least partially to overcome their racial prejudices and to concentrate on fighting . . . those in control of the southern political system." And the solid support that blacks demonstrated toward independent candidates evidenced a recognition by African Americans that the anti-Redeemer movements represented a genuine alternative to the racial politics embraced by both major parties.[16]

African Americans played a central role in the life of the insurgent movements. Particularly in northern Alabama, where between 1877 and 1882 the Greenback-Labor Party emerged as "the strongest advocates of the rights of blacks" in the Deep South, African Americans "figured prominently, sometimes dominantly" in party life.[17] The Alabama Greenbacks' "most charismatic leader," the black coal miner Willis Johnson Thomas, became so popular "that predominantly white clubs invited him to speak, and his revivalistic fervor resulted in interracial meetings." The colored Greenback Club at Jefferson Mines, of which Thomas was the "leading spirit," "had the best order and held the most regular meetings of any club" in Alabama, according to a prominent white party organizer, who added that he "never thought such work would ever be carried on by colored people." Local Democrats feared that "if we let this nigger alone he will ruin our whole State" and fretted that the rise of the Greenbacks threatened to overturn the delicate racial hierarchy that Redemption had only recently restored. "Three years ago," one dejected Democrat complained after a brush with Thomas in 1878, "if a negro dared to say anything about politics, or public speaking, or sitting on a jury . . . he would be driven out of the county, or shot, or hung in the woods. . . . *Now white people are backing them in doing such things.*" That the party's opponents found this latter aspect of anti-Redeemer activism so troubling is significant: reports of an imminent collapse of the color line were no doubt exaggerated, but the racial egalitarianism exhibited in the Greenback insurgency contrasted dramatically with the fierce racial hostility taking hold of the South.[18]

That the anti-Redeemer insurgencies established their most formidable strongholds in the mining camps of the Birmingham district is itself significant. Development of the coal region concentrated large numbers of black and white miners in areas already noted for defying the power of downstate elites and brought a cohesion to such sentiments that could not be matched in the hill-county districts of Georgia or Mississippi. Concentrated in the coal camps, where most of them worked as full-time miners, Greenback organizers circulated throughout the surrounding countryside, organizing local farmers into party clubs. The distinctly proletarian character of the northern Alabama Greenbacks imbued the organization with a bread-and-butter pragmatism that was lacking elsewhere. The local party program reflected not only the national organization's general concerns over the growing haughtiness of capital or the currency question but also promised concrete measures to redress grievances articulated at the local level by area miners. The party railed against the convict lease system and low wages and demanded an end to payment in company scrip. It gave voice to miners' complaints that they were being cheated through shortweighing of coal and pledged itself to work for more stringent safety laws and a rigorous system of mine inspection.[19]

With the decline of the Greenbacks after 1882, the Knights of Labor took up the mantle of coalfield interracialism. Like the Greenback clubs (having developed out of the remnants of the Greenback-Labor Party), the Knights leavened their agitation with a racial egalitarianism that was exceptional by contemporary standards. "We need a federation of men and women, white and black," Thomas Wisdom admonished an 1895 Labor Day gathering in Birmingham. "You [white] southerners may spurn the blacks but if you continue this the time may come when they will make you eat humble pie. . . . You strike and the capitalists will get them in your place." Like their precursor, local Knights assemblies operated as halfway houses between political parties and full-fledged trade unions, combining broad class demands with concrete agitation around local workplace grievances.[20]

Interestingly, it was their ambivalence over the propriety of direct action that opened a serious rift between the Knights of Labor and district miners. Although rank-and-file activists played a significant role in shaping the agenda pursued by local assemblies, important differences developed between these militants and state and national leaders. One major source of friction was the national organization's acute aversion to industrial action of any sort. Favoring arbitration as a means of settling grievances, the formal role played by the Knights' leadership during confrontations between miners and coal operators during this period "wavered between aloofness and active involvement," with tensions coming to a head in 1887 during a tense standoff between thirteen hundred miners and management at the Patton mines in Walker County. The Patton strikers, who had elected an autonomous strike committee to oversee their efforts, grew increasingly frustrated with attempts by the Knights' state master workman Nicholas Stack to impose an arbitrated settlement upon them. When miners rejected the offer negotiated on their behalf, Stack threatened to revoke their charter, an action that ultimately backfired, provoking ridicule and even mutiny from miners throughout the district. Recalcitrant Knights throughout Walker County and elsewhere denounced "in bitter terms the action of the executive board, and especially myself," recalled Stack, and the Coalburg (Jefferson County) local circulated an appeal calling upon miners throughout the district to defy their leadership and come to the material aid of the strikers. As Letwin has noted, the Patton strike heralded the decline of the nonconfrontational, hybrid political-industrial organizations that had dominated the coalfields and ushered in the rise of militant industrial unionism committed to direct action.[21]

It was this impasse, developing against a backdrop of increasing aggressiveness on the part of area operators, that provided the context for organization of the United Mine Workers in Alabama—initially (from the early 1890s) as a statewide organization and later as a district affiliated with the national UMW. Historians have debated whether the union's exceptional openness to the organization of black miners during this period derived from the union's "evangelical egalitarianism" or from more pragmatic imperatives. Detractors of the "new southern labor history," for example, have asserted that "Birmingham unions [including the UMW] organized blacks in order to control them." The most prominent exponent of this position, Herbert Hill, insists that blacks were admitted into unions "because it was in the interest of whites to do so." Moreover, he argues, "exploited white workers, through their unions, derived benefits from the greater and different oppression suffered by blacks."[22]

Hill's assertion contains an important kernel of truth. Like the pragmatic interracialism embraced by a minority of workers in the craft unions, coalfield interracialism was the product of restricted options. It was the operators, after all, and not white miners who controlled the racial composition of the workforce, and clearly low labor costs (and not the preservation of white racial solidarity) were their primary obsession. African Americans therefore comprised a large proportion of the mine workforce from the beginning: by 1900 they constituted the largest ethnic component in the coalfields and at the outbreak of the First World War formed an absolute majority. Interracial cooperation was therefore, in one sense, forced upon white miners for whom, as Letwin has noted, "the notion . . . of a racially defined territory at the workplace had little opportunity to take root." In contrast to the skilled trades, where the overwhelming predominance of whites made it possible to enforce an exclusionary policy, white coal miners—even if they were so disposed—lacked the power to purge blacks from the industry.[23]

The persistence of coalfield unionism in the face of sustained, calculated racial provocation suggests that something more substantial than white self-interest held black and white miners together, however. Hill's thesis presumes a static, undifferentiated racism among white miners and, perhaps more egregiously, stoic passivity on the part of African Americans. Black miners appear in the revisionist accounts not as the subjects of their history but as the mere pawns of white co-workers. The historical record suggests otherwise, however: black and white miners came together on terms over which neither exercised much control and in a society that provided few examples of extended racial détente, let alone cooperation. That the hostility so pervasive in every other aspect of southern life found its reflection in working-class organization is no doubt true, though unremarkable. As a *Labor Advocate* correspondent put it cogently, "It isn't to be wondered [that] there is this feeling against the colored man among some trade unionists, because there are so many eminently respectable citizens in the same communities who keep them out of their churches and all other organizations." He considered it "hardly a square deal to hold against the labor union the charge of discrimination, when all through the South . . . these folks in the supposedly better classes are doing the same thing." The penetration of a white supremacist sensibility into the ranks of organized labor is even less striking when one considers that leading employers, including especially district coal operators, consciously pursued a policy calculated to provoke racial antagonism.[24]

Though it is difficult to determine whether they did so out of a rudimentary sense of solidarity with black miners or an appreciation of the futility of exclusion, white miners seem never to have seriously entertained barring black miners from work in the mines.[25] Many did, however, embrace a variant of pragmatic interracialism that allowed them to reconcile deeply embedded race prejudice with assent to a tactical alliance with black miners. Even among those who accepted at face value the organic logic of white supremacy, many found themselves disoriented by the mine owners' violation of the very credo that they fought so hard to uphold during periods of industrial strife.

When, for example, Governor Comer boasted before a Democratic rally in 1910 that his aggressive strike intervention had "delivered" Alabama from "negro government" and "negro dominance" reminiscent of the "black republican days of the reconstruction period," his remarks elicited harsh condemnation in the pages of the *Labor Advocate*. Its editors condemned him as "a hypocritical mugwump, a monarchist, an impostor, and a leech on the body politic." The operators' increasingly transparent attempts to displace white mine labor with black struck white miners of the pragmatic persua-

sion as an infraction against the shibboleth of race solidarity. "White supremacy [Comer] certainly stands for," the former District 20 official R. A. Statham
complained after the speech. "This is proved by the fact that he does not want
anything but negroes to labor for him. . . . We shall soon bid you good bye,
governor, . . . and you can then devote your time to your negro colony and
every now and then, when you can spare the time, come down and make a
speech on 'White Supremacy.' It will help to open the people's eyes as to what
it takes to constitute white supremacy."[26]

Statham was not rejecting white supremacy itself so much as calling upon
employers to fulfill their "obligations" to less fortunate members of the white
race. After noting that a recent advertisement in the Birmingham papers had
added the proviso "MUST BE COLORED" to their list of qualifications for
employment in the mines, he protested that "those who seek employment
of the negro in preference to others should not be recognized as of a superi-
or race" since "it shows they are not in sympathy with their own kind and
that they are bloodsuckers of the worst type." "Talk about equal rights and
white supremacy," he complained. "It is all bosh when it comes to employing labor."[27]

Like many white miners, Statham was capable of genuine sympathy and
outrage at the treatment accorded black miners. "There are very few who do
not try to pull [the Negro's] leg for all they can," he conceded. "They treat
him worse than if he was a slave." Nor were they completely unaware that
displacement had as one of its aims the fomenting of antagonism between
black and white miners. Recounting the career of Henry F. DeBardeleben, the
architect of the "Negro Eden," Statham charged him with "growing rich"
from a career as a "union hater [who] never left an opportunity go by to cause
discension [sic] between white and colored labor." "We expect to see a lot of
[race-baiting] in the papers and we expect to hear the old saw, North! Yan-
kees, and Foreigners! And any old song to cause friction that will tend to keep
Alabama miners from uniting again."[28]

Statham's remarks represent an important strand of white working-class
opinion, which perceived—often accurately—the inequities arising out of
industrial exploitation but was too deeply ensconced within the white supremacist tradition to entertain the possibility of a clean break from racial
politics. One of the chief obstacles to such a development was their loyalty
to the Democratic Party. Unlike the period following Reconstruction, when
miners of both races had been attracted to the various independent formations, the politically stagnant years after the turn of the century offered little
room for organized electoral dissent for whites and none for blacks, who were
disfranchised in 1901. Except for the small Socialist Party, which retained a

respectable following among miners of both races, formal politics was reduced to a contest between varying shades of white supremacist opinion, in which issues of concern to labor were relegated to minor importance.

Given the powerful impediments to interracialism, however, what seems remarkable is not so much the level of animosity between black and white miners as their ability, through the experience of interracial cooperation, to begin to question certain fundamental aspects of southern racial protocol. While Hill and others have emphasized the ideological hold of racism on white workers, it is worth reiterating that there were many points around which the material interests of black and white miners converged.[29] Opposition to the convict lease system, resentment at the inequities associated with subcontracting, frustration with low wages and the company store, concerns over mine safety and the shortweighing of coal, the intrusion of the operators' authority into the most intimate aspects of family life: it was around these tangible concerns rather than racial sentiment in the abstract that black and white miners came together in a practical alliance.

Though the notion that white miners "benefited" from the oppression of blacks seems directly contradicted by the Alabama experience, it is true that black and white miners frequently brought distinct and sometimes conflicting priorities and concerns to their union activism. Convict miners, for example, were overwhelmingly black, and black free miners, many of whom had served their "apprenticeships" in the convict camps, seem to have felt more acutely a commitment to abolishing the system. But free white miners likewise had no stake in—and much to lose by—its continuation and were frequently led, in the course of defending their own material interests, to understand and condemn the racial injustice that was being carried out under the direction of the state. "There is no reason that a negro who steals a few chickens should have his hands cut off by working for a [coal company] for this offense," a correspondent reminded readers of the *Labor Advocate*. "There is no reason a white man should have his legs cut off in the convict mines . . . because he steals a ride on the train."[30]

Though blacks felt its effects more directly than white miners, the subcontracting system was another frequent source of complaint. A relatively small number of (mostly) white contractors benefited directly from the harsh exploitation that the system encouraged, but for the vast majority of whites subcontracting held in place an arrangement that guaranteed low wages and pitted independent miners against the gangs of low-paid laborers being worked by mine entrepreneurs. Miners "widely detested" the system, insisting that it violated the principle of "equal pay for equal work," and the UMW made its abolition a prominent demand in negotiations with mine management. Prior to the system's resurrection after the 1904 strike, the union had forced TCI to abandon subcontracting and replace it with a "buddy system," under which two partners worked underground at equal pay. Some years later, in an address before black strikers at Oakman, the District 20 president J. R. Kennamer acknowledged that "it may be pretty bad [for black miners] in some sections yet" but recalled that under the subcontract system "the negro done the work and the white man got the pay." "I can remember the time when the white man got a contract . . . and got two or three colored fellows to labor for him, and he went down and sat down and pulled off his coat . . . and says, 'Boys, lets run this car down,' . . . and that was your job." This system prevailed, Kennamer contended, until "the UMW came along [and said], 'Let us have an equal justice to all. Let the negro have a chance to dig coal if he is able to dig coal.'"[31]

The pervasiveness of racism among white southerners during this period ruled out any easy extension of job unity to more intimate aspects of social life in the mining district, and neither the UMW's constructive influence nor close personal relations developed on the job could transform the coal camps into sanctuaries from white supremacy. "Even when human sympathy and friendship drew people together," Edward Ayers has noted, "the rituals of Southern race relations constrained and distorted . . . feelings." A white UMW member expressing his appreciation to a dozen "colored boys" at Lehigh who, "as hard as times are," had "show[n] their friendship and loyalty to their mining camp [base]ball club" by riding into Birmingham to root for its white team, assured them that "we white folks . . . will not forget it" but apparently saw no contradiction in writing a month later to blame a fatal mining accident at Mulga mine on the company's employment of a black motorman. "Probably if they had a white man for motorman," he speculated, the accident would not have occurred. Two years later the same correspondent wrote to complain that "under conditions existing today in the coal mines of Alabama I am sinking fast to the level of the shiftless Negro." "For myself," he concluded in language redolent with the racial belligerence engulfing the South, "I don't care whether the Negro ever gets any better conditions if it will not effect [sic] me."[32]

Despite the persistence of such attitudes, the isolation of the mining communities, the ever-present (and nondiscriminatory) danger intrinsic to coal mining, and the experience of combined struggle against the operators did draw blacks and whites together, laying the basis for an interracial camaraderie unique in the region. "With one or two exceptions," the United States Immigration Commission reported in 1911, "there has been very little friction among the races employed." At a time when lynchings and similar racial outrages were common throughout the South, Letwin notes, "one is struck by the apparent infrequency of such open tensions" in the coal camps. Though most companies formally upheld segregation in their living and working arrangements, there were many areas, both under and above ground, where the lives of black and white miners and their families intersected. "You worked shoulder to shoulder with a man," recalled a black TCI miner. "You['d] go over and help him push his cars and he'd come over and help you push your cars up and they didn't [care]. . . . When they went into the mines they was all miners and they didn't hold nothing agin' nobody." A section foreman from the Woodward Coal and Iron Company concurred. "We never had no [racial] trouble," he told interviewers. "If one of them hurt, all of them hurt."[33]

The unpredictability of mine disasters and the daily hardship encountered during slack periods and strikes galvanized a similar affinity between miners' wives, a sense of commonality that could likewise puncture if not transcend the color bar. Dera Davis, the wife of a white subcontractor at Sayreton, told interviewers of having been summoned to the Dogwood mine by a shrill whistle and the "sound of terrible shrieks." She arrived to find "bodies burnt . . . so that you couldn't tell whether they was black or white." After mine officials informed a distraught crowd of women and children that no blacks had been working the section of the mine where the explosion had occurred, Davis recalled hearing a black woman exclaim that she felt "just as sorry for these folks as if they was my own color." The same elemental solidarity became evident during prolonged strikes, as documented in the rough prose composed by Rossia Cooker, a self-described "striking union woman" from the UMW local at America, Alabama, who called on her "friends both white and colored" to "fight on" until "scabbing [would] be no more."[34]

Close personal relations across the color line alone could not bridge the chasm that had divided black and white southerners for centuries. The remarkable tradition pioneered by the anti-Redeemer movements provided an important precedent for interracial cooperation. But the key element in breaking down racial divisions in the mines was provided—or imposed— by the coal operators. With trepidation and great reluctance at times, white miners came to discover that, far from benefiting from black oppression, they could not hope to improve their own situation without simultaneously fighting to alleviate the appalling conditions under which black miners lived and worked. Even relatively conservative white union members approved of the UMW's making "no distinction in its members for creed, color or nationality." "*We are like the coal operators,*" wrote Statham. "They employ all kinds of labor and we take all kinds into the union. . . . We don't believe because a man's skin is black he should perform labor any cheaper than a white skinned man." "There is only one way for us to better our condition, and that is by bettering the Negro's condition," a white Searles miner wrote, adapting a metaphor popularized by the Populists and revived in a different context by Booker T. Washington: "It is a fact that we can not rise and leave him in the ditch."[35]

The most emphatic declarations of this sentiment reflected southern racial sensibilities even as they challenged the status quo, combining incipient racial egalitarianism with a paternalist logic of their own. "In order to save the industrial life of the white man," the correspondent quoted above continued, "we must lay aside prejudice and face the situation like white men with backbone and take the negro by the hand and lift him out of the damnable industrial pit and make him a man, and thus by saving him we will save ourselves, our wives and children."[36]

The patronizing tone with which white miners and union officials frequently wrapped their embrace of interracialism, though confirming Ayers's dictum about the capacity of race prejudice to intrude upon even the most benign relationships between blacks and whites, must nevertheless be distinguished from the paternalism of the operators. While in their more candid moments the latter admitted that it was the vulnerability of black workers that they found so attractive and that their aim was to wield that defenselessness as a means of reducing the costs of mine labor generally, white miners approached their relations with blacks from a very different perspective. Mired as they were in the heritage of white supremacy, white miners nevertheless realized that it was only by improving the situation of black workers that they could better their own lot. While fundamental inequality, in terms of wealth and power, contorted relations between even the most benevolent of operators and their black employees, black and white miners dealt with each other from positions of relative equality. Despite its imperfections and its tendency to accommodate rather than directly challenge racial bigotry, interracial unionism represented a sharp challenge to the southern social order and posed a direct, intrinsic threat to the foundations of Jim Crow.

Some white miners, it should be emphasized, traveled far beyond the conservative pragmatism expounded by Statham. From an elementary recognition of the necessity of aiding the organization of black miners, some whites became convinced that they would have to directly confront the racial abuse suffered by blacks in order to mount a successful challenge to the power of their employers and make interracialism effective. After objecting that black miners were, "in many instances, treated worse than convict[s] by the superintendents," a white UMW veteran urged renewed efforts to recruit black miners to the union. "This should be done," he urged, "not only in self-defense, but for the sake of humanity, as the Negro is worthy of being saved from the industrial and moral death he is dying."[37]

The distance between this sentiment and that of the conservative pragmatists can be gleaned from the corresponding proposals for curtailing the abuse meted out to black miners by company "shack rousters." Although Statham spoke harshly of the shack rouster, who swaggered about the camps "carr[ying] a few pistols, a baseball bat, locks and chains," while "giv[ing] orders in true slave-driver style," he stopped short of urging that the UMW take concrete action to prevent such outrages. Instead, with the bravado that comes easily to someone exempt from such treatment, he warned that "if the day comes when a shack rouster comes around my home and asks me why I am not at work, he had better shoot at the same time, or I will sure violate one of Moses' commandments." The problem, Statham seemed to imply, was that black miners lacked the manhood necessary to defend themselves. The UMW veteran miner quoted above viewed the problem quite differently. "The operators do not send the 'shack rouster' to any white man's home to drive him out to work, sick or well, but day by day the reigns are drawn tighter . . . and if we continue to hold our peace the day will soon come when we will be catching just as much hell as the Negro." "The time to act," he implored readers of the *Labor Advocate* in 1916, "is now."[38]

For their part, black miners did not simply wait on the aroused sympathy of their white companions to press their grievances. Drawing on a rich tradition of black working-class activism, with roots in the Greenback-Labor Party and perhaps even further back into the Union Leagues and the radical wing of the Republican Party, black miners proved themselves adept at fusing workplace militancy with racial pride. Though highly visible, articulate black militants like Willis Johnson Thomas were exceptional, they represented a broader strand of black working-class militancy that must be taken into account in any balanced assessment of the complex record of interracialism.

The working lives of even the most prominent black unionists can only be partially reconstructed. Silas Brooks, the District 20 vice president at the turn of the century, earned a footnote in the history of Alabama unionism when he stood up to national UMW officials against their collaboration (in the form of "card exchange") with "lily-white" craft unions. Brooks's successor, B. L. Greer, was severely beaten by company thugs along with a white UMW organizer at Horse Creek in 1903. A labor journalist reporting on Greer's remarks before an audience of black miners at Sayreton three years later observed that the "colored people present . . . seemed to drink in everything he said, and Greer is capable of saying many good things." Will Millin, who emerged as a local leader at Dolomite during the 1908 strike while "advocating extreme measures against the [Woodward] company," insisted that "all miners and coke workers quit work and join the union." Millin was arrested for his audacity, taken from jail, and lynched by a party that included two company officials (an outrage that provoked fierce, armed retaliation by a racially mixed group of union miners).[39]

Two black UMW militants stand out in the post-1908 period. Walter Jones, who succeeded Greer as District 20 vice president, was "born" to the union at Aldrich, where his Virginia-born father had found work after emancipation. A hardshell Baptist whose brother, a preacher and union stalwart, had been shot during the 1908 strike, Jones was revered among black miners for his fearlessness and oratorical genius. When vigilantes attempted to dissuade him from speaking at a mass meeting by threatening assassination, Jones had responded calmly, "Well, that's what you're going to have to do. I don't believe a man here got nerve enough to shoot me down off this platform. If it takes death to settle it, I'm willing to go." "He could give those companies the devil," a former TCI employee recalled admiringly, "and people would come from miles just to hear him." Jones's contemporary, Joe Sorsby, would provoke the wrath of the ACOA for the prominent role he played in the 1920 strike.[40]

Less well known are the lives of black rank-and-file miners. Perhaps the most telling evidence of the depth of black militancy is contained in the moving record left by the Brookside miner Dan DeJarnette who, despite having been blacklisted for nearly twelve years after his role in the 1908 strike, described himself as having "not one drop of scabbing blood in me." DeJarnette, "sure enough barefoot" and nearly naked "with my last pants on and they . . . all torned up," wrote District 20 officials several months into the 1920 strike offering to travel south to the Black Belt, where operators were recruiting strikebreakers. "I can go through these counties and explain to these poor eanorance [sic] farming people the facts of it all and stop them," he assured UMW officials. "It is the Negro . . . they are hitting on, my race."[41]

The fierce loyalty of these black union militants calls into question the operators' projection of blacks as a bulwark against labor agitation. How do we reconcile the mine owners' stated preference for "cheap, docile negro labor" with evidence that blacks played a leading role in making the UMW a viable force in the Birmingham district? Interestingly, operators and seasoned miners seem to have distinguished between experienced black miners and "green negroes," who had only recently arrived in the district. The operators' encouragement of a steady influx of newcomers from the Black Belt not only provided them with an abundant labor supply but with a large pool of

The headstone erected by UMW District 20 at the grave of Walter Jones and his wife, Alice. Jones's father had been born a slave and moved to Alabama from Virginia after emancipation, where he became active in early organizing efforts. Jones and his brother, the miner/minister Fred C. Jones, were officers in the UMW local at Aldrich. Walter was revered among black miners in the district as a fearless union militant and a powerful orator. (Courtesy of Constance Jones-Price and the Birmingham Public Library Archives, Birmingham, Ala.)

tractable laborers unlikely to risk their hard-won new prospects by joining a trade union. The numerous examples of attacks on black strikebreakers by members of their own race or of combined black and white armed assaults on private and public law enforcement officials suggests that loyalty to the UMW could often submerge racial divisions.

The problem, from the operators' perspective, arose as soon as new arrivals had adapted sufficiently to the coal district to realize how far their new surroundings fell short of their expectations. Within five years of the 1908 defeat, miners of both races were commenting on the near-unanimity among black miners to reorganize the UMW. When in 1913 the prominent black antiunionist P. Colfax Rameau published a letter lauding the mine owners' benevolence in the *Birmingham News,* a white miner responded that "[Rameau] dare not come here and tell these negroes that the union was not their best friend [or] they would make it hot as Hades for him." "Almost every negro miner in the district has sworn never to give up the struggle until we have planted the banner of unionism in every camp," a black miner assured readers of the *UMW Journal* in 1916. "I know the sentiment of the Negro miners, as we very often meet and talk," a black Republic miner explained, adding that they "almost to a man recognize that they should be organized." Trusting that it would bring "joy to every white miner to know that the eyes of our brother in black skin are being open [sic] to the conditions under which we are working," a group of "Colored Miners" from Bessie and Palos mines expressed their resolve that "the main issue is we must have a union in Alabama at any cost."[42]

Significantly, every one of these expressions of black miners' support for the UMW contained explicit declarations that in return for their loyalty they expected a square deal from the union. The *UMW Journal* correspondent compared his people to "the children of Israel" who "prayed for deliverance." Like the Jews, black miners were "praying for the dawning of a bright, new day when God . . . will send us a leader . . . whom we can wholly confide in." "*If the UMW will stand in those shoes,*" he vowed, "we will hail their coming with unexpressable joy." That his religious imagery conveyed a temporal meaning is evident in his suggestion as to how the UMW might fulfill its role: "With the union's banner floating over the different camps of old Alabama, [the shack rouster] would have to hunt more congenial climes, as the camp life would be too warm a place for him." The Bessie and Palos miners urged union officials to reach out to black preachers, including Rameau, and "prove to [them] *that the Negro will get a square deal at the hands of the UMWA.*" After outlining the harsh conditions under which black miners labored, the Republic correspondent "ask[ed] on behalf of hundreds of Negro miners" if "the Union can remedy these damnable conditions and make camp life worth living." "We want an organization," he insisted, "*but not at any cost.*"[43]

Their keen insistence on a quid pro quo reveals a level of confidence among black miners that one would not expect if they lacked real influence over union affairs. Despite internal and external resistance to black equality, their sheer numbers made it impossible to deny black miners a voice in the UMW. Constituting an increasing proportion of the mine workforce (and after 1900 a majority), blacks exercised considerable power in steering the union toward racial egalitarianism. The antidiscrimination clause negotiated in the 1898 contract may well have been the product of negotiations between black miners and white union officials over an equitable solution to the problem of subcontracting and was almost certainly the result of pressure applied by black unionists. The District 20 vice president Silas Brooks's intervention in the Alabama UMW's 1900 convention provides a powerful illustration of the lengths to which black miners would go to defend themselves against racial abuse and their confidence about pressing the issue within the union. "Even [national UMW] President John Mitchell," Ronald Lewis recounts, "could not persuade black delegates to vote [for a resolution] which called upon delegates to support a proposal of the Birmingham Trades Council that miners patronize only union workmen." Brooks objected on the grounds that "the Council discriminated against Afro-Americans, and he persuaded the convention to reject the resolution." B. L. Greer adopted the same position when the issue came up at the UMW's national convention four years later. "I understand that a man of my race cannot belong to [the blacksmiths'] union," he argued. "We do not want to accept [them] when a part of our membership could not be taken into their union on transfer cards."

Testimony from mine management and white union officials suggests that, particularly after 1916, black miners came increasingly to view the UMW as a vehicle for racial as well as economic deliverance. Lurid complaints on the part of mine owners that black union organizers were "promis[ing black unionists] equal vote and equal marriage laws," though certainly exaggerated, seem to confirm that the boundary between racial and economic demands was often unclear and that black miners were comfortable with the confusion.[44]

The Jim Crow South afforded little room for interracial collaboration of any kind; the organization of black and white workers for the purpose of countering the power of the region's leading industrial interests constituted an extremely precarious undertaking. District 20 faced implacable and near-universal hostility, much of it centering on the union's commitment to interracialism. The pressures of operating in this hostile environment imposed a certain ambiguity upon the union's racial policy, an ambiguity that existed independent of the attitude of white miners and that many black miners apparently accepted. The clearest example of this is the so-called UMW formula, a method by which the Alabama UMW elected local and district leadership. The formula represented an explicit concession to white supremacy: of seven leadership positions in racially mixed miners' locals, four went to whites, while three (including the position of vice president) were reserved for blacks. This arrangement, though obviously calculated to mollify racists within and outside the union, was nevertheless advanced by the standards of the contemporary labor movement and virtually unique among southern institutions. At its best the formula provided a means through which district leadership "balanced equal rights and full participation for its black members against the outward signs of white supremacy."[45]

In black-majority union locals, district officials defended the formula's violation of democracy as a necessary concession in a society that would barely tolerate racially mixed organizations, let alone black-led ones. And internal union debates seem to indicate that, for the most part, both black and white miners accepted the logic of this position. When the matter was discussed at District 20's 1904 convention, a white delegate rose to argue against popular election of union officials, contending that such a course would produce "race troubles." "Electing a president from among the colored brethren," he warned, "would mean the destruction of the organization in the South." This objection was echoed by a black delegate from Blocton, who successfully urged black miners to decline the popular vote, "even though with it negroes would probably capture all the offices in the district organization." The obvious inequity embodied in the UMW formula was likely viewed by black miners as a sacrifice to be borne until such time as it could be overturned—never as an equitable solution.[46]

In locals with a heavy black majority, tensions over representation surfaced occasionally. The *Birmingham Reporter* related an incident that allegedly occurred at Praco Mines in July 1917, after a black majority elected one of their own as local president. According to a Praco correspondent, white miners approached the newly elected candidate and "advised him to quietly and submissively resign and let one of the white brethren have it." In the view of the editors, the incident confirmed the folly of "Negroes associating with white folks." Whether or not such incidents occurred on a regular basis, it is obvious that the ambiguity imposed upon the UMW from the outside would have reinforced the efforts of those determined to keep black miners in their place within the union.[47]

On the whole, however, relations between black and white miners seem to confirm Ronald Lewis's assertion that "whatever local discrimination blacks encountered in their dealing with white miners, in the context of southern society the UMWA was the most progressive force in their lives." Black miners were not mere observers of union affairs but appreciated their indispensability to the UMW's efforts and applied their leverage to demand fair treatment. In a society thoroughly steeped in white supremacy and prone to violent assertion of that heritage, black and white Alabama miners drew upon a tradition of plebeian and working-class interracialism to forge a culture of resistance against the power of the coal operators. In the process of doing so they elicited the scorn of respectable white southern society. Not only Birmingham's business elite but even the more conservative sections of the organized labor movement distanced themselves from the UMW's interracialism. "[I'm] told that some negroes make good miners," an engineer on the Louisville and Nashville Railroad unwittingly informed ACOA detectives, "though when he gets to be a union man he generally thinks he is as good as a white man." Buried for years following defeat of the 1908 strike, coalfield interracialism was about to be tested again in the new conditions of wartime Alabama.[48]

5. War, Migration, and the Revival of Coalfield Militancy

IN EARLY 1914, when Birmingham-area coal operators convened to review their activities over the preceding year, they could look back upon the half-decade that had elapsed since the 1908 strike with a sense of having successfully begun the transformation of the mineral district. Despite troubling evidence that the coal industry was entering a downturn, operators congratulated themselves on having expanded Alabama's mine workforce by two thousand men in only a year. Under their sagacious administration, ACOA spokesmen claimed, miners' wages had increased, living conditions had been dramatically improved, and sanitation at the camps was now "almost perfect." The long-term commitment that leading employers had made to welfare capitalism seemed at last to be bearing fruit: over and above their material achievements, operators derived their greatest satisfaction from having "inculcated the idea in the minds of our employees that we are their friends, and that they and we are mutually dependent, one upon the other." The "peculiar control" that federal officials had ascribed to Birmingham operators several years earlier seemed to have become fixed as a permanent feature of labor relations in the Alabama coalfields. "We have fully demonstrated the axiom that 'in unity there is strength,'" ACOA officials noted confidently, with the result that "not a ripple of discontent or disorder [had] marred" day-to-day operation of the mines over the previous year.[1]

Developments over the next several years would demonstrate that the operators' faith in the unwavering loyalty of their workforce was unfounded, however. The carefully guarded tranquillity then prevailing throughout the district—welcomed by mine owners as the chief dividend accruing to them from their embrace of welfare capitalism—concealed but could not eliminate seething bitterness prevalent among rank-and-file miners. "It matters not where I see men from, they all say the same thing," one miner wrote the *Labor Advocate* in September 1913. "'When are we going to be organized again?'" "You let some one get out among these Alabama miners with the power of speech," he predicted, and "you would think there had been an old-time camp meeting and everybody joined the church." A national UMW delegation sent south to assess the possibilities for relaunching the union during the same year concurred, acknowledging that "sentiment for organization could not be any stronger among the miners than it is now." Though convinced that "practically every free miner in the field would respond to a strike call," delegates advised against such a course of action until the national organization could shoulder the financial burden of sustaining a long drawn-out strike. "It is a question of money, and money alone in this district," they informed the UMW president John P. White, since "the coal companies are going to give battle before they will submit to having their employees organized."[2]

While the sober appraisal submitted by UMW field-workers reflected real sentiment for unionization in the district, its authors clearly underestimated the difficulties involved in translating such aspirations into open defiance of the operators' authority. When UMW officials held a mass meeting in Birmingham to discuss reorganization in March 1913, local supporters expressed bitter disappointment that "the house was not fuller." While acknowledging that many miners had stayed away from fear of victimization, district unionists nevertheless considered the low turnout a "discredit [to] the so-called union men of Birmingham."[3]

The employers' solid hegemony after 1908 had greatly reduced the possibilities for collective action. Their program of material improvements, while it never delivered the idyllic conditions depicted in ACOA propaganda, won the acquiescence of a substantial proportion of the mine workforce. The continual turnover in mine labor posed a serious impediment to any efforts to establish permanent organization. And the operators' reluctance to part with the coercive element of their labor relations strategy—the blacklists, the routine application of corporal punishment, the ubiquitous deployment of company guards and undercover detectives—made open expression of discontent an extremely precarious affair. Conditions practically ruled out sustained, coordinated protest, and the few minor confrontations that did develop between miners and their employers during the period between 1908 and 1916—as in the disputes over application of mine safety legislation and the appointment of checkweighmen—remained localized and episodic in nature.

Despite these serious obstacles, however, over the next three years the balance of power in the Alabama coalfields shifted dramatically. The swagger-

ing optimism that permeated the outlook of ACOA members in 1914 gave
way to despondency and outright panic as the "peculiar control" they had
enjoyed since 1908 increasingly slipped from their grasp. The "old-time camp
meeting" enthusiasm, which even at the height of the operators' hegemony
UMW stalwarts confidently expected in response to a sustained reorganiz-
ing effort, had spread through the district like wildfire by mid-1917. From a
dismal low of only twenty-eight members and five weak union locals across
the coalfields in 1915, the UMW launched a reorganization in the spring of
1917. Mass meetings of up to three thousand miners were held at Brookside
and elsewhere throughout the Warrior coalfield. Rallies at Jasper, Wylam,
Seymour, Maylene, Sumter, and Blocton drew large turnouts and were everywhere marked by "revival camp enthusiasm." Within two months, District 20 officials reported having organized 128 new locals, with twenty-three
thousand of twenty-five thousand miners in the district now holding union
cards. Over a period of several months the UMW had managed to achieve
what would have been impossible several years earlier: "almost total organization of the coal mines."[4]

This dramatic reversal of fortune was the result neither of a slackening of
vigilance on the part of area operators nor of newfound enthusiasm among
union miners but of a series of much more fundamental transformations set
in motion by the advent of World War I. The sea-change in coalfield labor
relations exposed the frailty of the operators' new order and revealed how
little it had ever actually relied upon the consent of its ostensible beneficiaries. Black and white Alabamians had endured—rather than welcomed—the
operators' unilateral authority in the mines, and when, under the weight of
extraordinary pressures brought on by the war, the mine owners' hold over
affairs began to weaken, miners took advantage of their disarray to register
their dissatisfaction.

For nearly four decades preceding the outbreak of war in Europe, Alabama
operators had enjoyed an unparalleled asset in their close proximity to the
Black Belt. The edifice of labor relations constructed by the ACOA after their
victory in 1908 had as its cornerstone a continuous supply of cheap, relatively
defenseless black labor. Not only the coal industry but the iron ore mines,
steel mills, and iron foundries were all heavily dependent upon a steady supply of black workers desperate enough to labor under conditions and at wages
that ensured their employers' competitive advantage. When new circumstances brought on by the war began to interrupt this supply and, later, to
inject a new sense of confidence and militancy among Alabama blacks, the
operators' expectations of a placid future were upended.

Well before formal U.S. entry into the war in April 1917, American industry had already begun to feel the impact of conflict in Europe. To the relief of many in the business community, increased war demand had by early 1915 begun to lift the national economy out of the slump it had experienced since 1913. The revival of industrial production posed difficulties of its own, however. War in Europe drained American employers of their main source of unskilled labor. Over a half-million European immigrants returned home to fight under their national flags, while the number of those entering the United States slowed to a trickle: from 1.2 million immigrants in 1914 their numbers declined drastically, to just over 320,000 a year later and less than 120,000 by 1918. Hard-pressed to keep pace with growing war demand, northern industrialists turned for the first time to southern blacks as "the only untapped source of common labor remaining in the country."[5]

As the growth of Birmingham's mineral district illustrates, black migration out of the Cotton Belt had been underway since the end of the Civil War, or since the moment freed slaves had secured their legal right to move on. Prior to World War I southern cities like Birmingham served as the final destinations for those seeking relief. In the new circumstances brought on by the war, however, the willingness of northern employers to pay transportation out of the South and their dispatching of labor agents throughout the region "offering two dollars, three dollars, four dollars and five dollars a day" to "southern negro workmen . . . averaging only one dollar to two dollars a day, and many of them less" meant that most of those deserting the cotton fields after 1915 had their sights set on Cleveland, Detroit, or Chicago rather than Birmingham or Atlanta.[6]

Historians have vigorously debated the relative weight of economic, political, and racial factors in spurring Black Belt laborers to desert the plantation districts. It is unlikely, however, that those swept up in the exodus spent much time splitting hairs over such matters. Cataloguing the list of grievances related to them by those taking part in the migration ("the pinch of poverty, want and hunger, the poverty of the educational system"), Urban League investigators reported a "pretty general desire to leave it all behind." The letters of migrants themselves illustrate that for many the abject poverty in which their lives seemed trapped was indissolubly bound up with their subordination under a regime committed to upholding white supremacy. "We are working people but we can't hardly live here," a frustrated black Louisianian complained to Urban League officials. "I would say more but . . . we have to lie low." "There's been so much killing of Negroes around here I didn't know when my turn might come," a Georgia native told reporters after reaching the North. From Shreveport came a query "as to how and where I can go to obtain better freedom and better pay for the balance of my life."

"Would work under any conditions to get away from this place," a black Texan informed editors of the *Chicago Defender*, "for I am working and throwing away my valuable time for nothing."[7]

Letters home from friends and relatives who had already emigrated reassured those tempted to leave the South that a new life of freedom and prosperity awaited them above the Mason-Dixon line. Southern blacks' somewhat charitable impression of the North as a refuge from racial hostility, rooted in an embellished recollection of its role in the Civil War, seemed confirmed in the impressions related by new arrivals to the North. "I should have been here twenty years ago," a newcomer to Chicago wrote his relatives in Hattiesburg, Mississippi. "I just begin to feel like a man." The black coal miner W. L. McMillen, an Alabama native recently arrived at Omar, West Virginia, sent a short note urging those he had left behind to head north. Coal operators there, he reported, were "prepared to take good care of new comers." "You can make one dollar far quicker than you can make twenty [cents] theaire [*sic*]," he assured friends. "The smallest wage is $2.00 per day on up to $4.00 so no way to miss a job heaire." Equally precious to McMillen was his right to cast a ballot. "I will vote for the President on the 11th of this mont[h]," he declared proudly. "Collered men tick stands just as good as white men here."[8]

The resolve of tens of thousands of black southerners to seek a better life in northern cities was steeled by the fact that for many the already difficult conditions with which they were familiar in the Black Belt declined further still from 1915 onward. Increased industrial demand in the North coincided with a particularly severe crisis in cotton production. While before 1911 Alabama's cotton harvest had expanded steadily, reaching a peak of nearly four million acres in that year, intensified competition on the world market combined with the domestic slump to force prices down and plunged the industry into crisis—one that sent ripples throughout the southern economy. Southern communities that had been relatively insulated from the industrial downturn felt so acutely in the North by 1914 "were bustling gaily along until suddenly there came this catapulting into the slough of despondency."[9]

Black tenants and sharecroppers found themselves pushed out of the plantation districts not only by the invisible hand of the market but by a series of devastating natural disasters as well. The cotton crop was ravaged for three consecutive planting seasons by the boll weevil, which first appeared in the Alabama Black Belt in 1912, and then by severe flooding in 1915 and 1916. During the summer of 1916, state officials informed the attorney general in Washington, "middle Alabama was visited with a terrible storm and the crops were destroyed," with the result that planters were "unable to furnish sustenance to the negro laborer on the farms." After conducting an investigation of conditions in southern Alabama, the federal attorney Alexander Pitts informed his superiors that the cotton crop in Dallas County had declined from sixty-seven thousand bales in 1914 to under five thousand two years later. "All of the surrounding counties are in the same condition," he observed, and "since this country only raises cotton and corn . . . the negroes have nothing to eat. The planters are not able to feed them and they are emigrating." Relief workers reported from the Cotton Belt in early 1917 the sight of hundreds of "hungry, half-naked, bare-footed poor Negroes huddled on the frozen ground, waiting their turn to get a little ration of meat and a peck of meal."[10]

From being relieved at the siphoning off of surplus labor during its initial phase, agricultural and industrial employers came to view the gathering exodus with increasing alarm. The *Montgomery Advertiser,* noting that "hundreds of Alabama Negroes, mostly from the Black Belt," had left the southern half of the state for "northern industrial centers" and the "coal mines around Birmingham," warned in August 1916 that it was a "mistake not to induce our Negroes to stay in Alabama." "If they go," editors worried, "where shall we get labor to take their places?" As the trickle of blacks out of the plantation districts threatened to become a stampede, African-American observers noted with some irony a "reversal of attitude on the part of whites toward Negroes." "The same press now so solicitous of the Negro's welfare," Urban League investigators remarked wryly, "has been industriously preaching for a quarter of a century the displacing of Negro labor on the farms, in the shops and the factories with white pauper labor from Europe." The reasoning behind the white South's reappraisal of race relations was spelled out most clearly in the *Macon (Georgia) Telegraph.* Frustrated that "everybody seems to be asleep about what is going on right under their noses," editors attempted to rouse white southerners to the peril that awaited the region if the exodus continued unhindered. "We must have the Negro in the South," they warned. "He has been with us so long that our whole industrial, commercial and agricultural structure has been built on a black foundation. It is the only labor we have; it is the best we have—if we lost it we [would] go bankrupt."[11]

Birmingham's industrial employers attempted to counter the flight of black labor out of the district in several ways. Company welfare programs developed after 1908 found a new sense of purpose in the drive to hold labor to the district, and welfare workers toured the district extolling the advantages of staying put.[12] One of the early casualties of the exodus was the operators' determination to hold the line against increased labor costs. Despite their long-standing aversion to raising wages, such a course seemed the most obvious means of holding their skilled labor. Wage rates began to rise throughout the district from well before formal U.S. involvement in the war. From an average of forty-seven-and-a-half cents per ton on Pratt Seam coal in 1913, wages climbed steadily by about ten cents per year, until district miners were averaging anywhere from eighty cents to a dollar per ton by December 1917. Explaining his extraordinary attempt to renegotiate prices with long-term clients of the DeBardeleben Coal Company in December 1916, the Sipsey mine supervisor Milton Fies complained that the exodus had confronted Birmingham operators with "conditions probably without parallel since the Civil War." "It has been necessary for us to offer inducements to our miners to prevent more of them from migrating northward," he explained to a New Orleans buyer, "where manufacturers of war munitions, machinery, [and] automobiles . . . are making very attractive offers to southern labor."

The changed conditions seemed to operators to have not only driven up the cost of labor but also to have reduced its efficiency. "Discipline has no effect," Fies complained bewilderedly, "because if you discharge a man he knows he can go to Bessemer and get a ticket to Cincinnati or some other place up North." "We have so completely lost discipline in the camp and in the mines," his superior, Henry T. DeBardeleben, commiserated, "that it is a hard matter to know one day what the output will be the next."[13]

On the whole, the material incentives that Birmingham operators were willing to extend fell far short of stemming the exodus. Birmingham wages remained higher than elsewhere in the South, as they had been for some time before the war, but considerably lower than those on offer in comparable northern cities. The black TCI labor agent Melvin Chisum[14] protested in a letter to the Tuskegee president Robert R. Moton that while the South was "getting its share of the National Pork Barrel, in every sort of a way," the region's employers were "bound and bent upon withholding all [they] can from the Negro toiler." Particularly worrying to Birmingham operators was the fact that skilled workers—black and white—were among those who seemed most inclined to head north. Conspicuously omitting any distinction between black and white emigrants, editors at the *Labor Advocate* advised that there existed "only one way to keep labor in the Birmingham district, and that is to pay the same wages that are being paid in other fields." They advised district coal operators that if they wanted "to secure the best workers in the country and wish the return of those workers ['both white and colored'] who have left this district by the thousands," they should consent to allowing the miners representation by the UMW. Otherwise, the *Advocate* reasoned, "all the hullabaloo and noise will not prevent them from going." The mixed composition of those departing the district was confirmed by

Survey magazine, which reported that southern whites "moved North in large numbers during the same time. Some of the causes moving the Negroes they have not felt; others they have."[15]

Equally significant and too often overlooked was the role that Birmingham played as a transit point between the Cotton Belt and the North. "There are negroes leaving here by the carload every day," came a report from Selma, "but they are all going to Birmingham and [being] reshipped from there to outside points." Perhaps a majority of black laborers leaving Alabama's Black Belt with their eyes set on Chicago ended up spending an extended stay in Birmingham; many found employment in the district's mines and mills, where they stayed for weeks or months, and occasionally they made the district their permanent home. Urban League investigators described Birmingham as an "assembling point for prospective migrants, who come and are made familiar with the industrial life which they are to meet when they reach the North."[16]

In the years directly preceding American involvement in World War I, the continuous flow of migrants into the district seems to have fueled a steady migration of black and white skilled workers north out of Birmingham. The rapid influx of unskilled migrants from the Cotton Belt after 1916 undermined the wages of experienced whites and blacks, who found themselves being aggressively recruited by coal operators from northern fields in West Virginia and Kentucky. The U.S. attorney for southern Alabama, Alex D. Pitts, concluded after a detailed investigation that "the great bulk of [the seventy-five thousand blacks who had left Dallas County] have been kept and are now working in and around Birmingham." "They were employed in Birmingham," he wrote, "because they could be employed cheaper than the negroes who were already there and that forced the negroes who were already there to leave." Noting that "this district has, for many years, been opposed to organized or union coal miners" and that "the operators have sought in every way to break up such unions," Pitts's counterpart in Birmingham confirmed that some of those fleeing the Black Belt, "in [their] desperation, worked very cheaply in the mines, and this put some good miners out of employment, . . . open[ing] up a fine field for the labor agents."[17]

At the same time that the exodus exerted pressure on the operators to raise wages, the flooding of the district with an army of impoverished laborers softened the blow, albeit at the cost of inducing experienced miners to head north. Ultimately, however, Birmingham industry was unable or unwilling to match the wages on offer in northern fields and began to draw upon more traditional methods to check the mobility of black labor. The TCI manager James Bowron recalled that Birmingham banks "refused to cash checks written by northerners and sent South to finance the 'darky's joyride.'" The Southern Metal Trades' Association issued a bulletin advising employers to "shut the barn door before the horses all get out," and a number of operators seem to have applied the journal's directive enthusiastically, as an incident related by the *Labor Advocate* in June 1917 reveals. At a mine located just outside city limits, "many of the colored population had put on their best Sunday clothes and had gathered at the depot intending to come to Birmingham to a big baptizing. Just before the arrival of the train *the superintendent and shack rouster* appeared upon the scene and forced all the negroes, without allowing them time to go home and put on their mining clothes, to go into the mines to work." Given such "intolerance," the editors reasoned, it was no wonder that "the colored people . . . enter their protests in the way they have." Until recently, one white miner charged, the operators had "depended upon fleecing the negro, getting the negro's labor for a song and making the negro sing it." The exodus was proof, however, that "the negro, having gotten tired of this way of doing things and having heard of better conditions in the north . . . is anxious for the free ride."[18]

Frustrated at their inability to halt migration out of the district, industrial employers looked to their allies in Birmingham's black middle class for assistance in inoculating black workers against the "northern fever." Although individuals like Oscar Adams ultimately assumed the role of apologists for their friends in industry, even the most dyed-in-the-wool accommodationists recognized in the unique wartime conditions an opportunity to advance the interests of black southerners. They embraced the logic spelled out by editors of one northern newspaper that the tumult "at least promise[d] an amelioration of the condition of the Negro laborer in the South." "To keep the Negro at home," newspapermen reasoned, "southern industry . . . will pay him a little more. . . . The Negro's opportunity will surely improve with the demand for him." Dr. J. H. Eason, the pastor of Birmingham's Jackson Street Baptist Church, feared that the "rush" that had "come among our people for the North" would develop into a "stampede," in which "many people would be hurt in its final results." Nevertheless he detected "some good in this great migration of Negroes out of the South." Similarly, after chronicling the flight of black labor out of neighboring Georgia and warning that the *Reporter* "[did] not agree that it is the best thing for so many of our good citizens to leave the Southland," Adams chided white southerners that "there is another deep seated trouble connected with this labor exodus." For the past thirty years, he protested, while being able to "read about civilization, hear about civilization and talk about civilization," southern blacks had been "forever hungering and thirsting for just a little of it, and not that amount that the Southern white man has designated as sufficient for his capacity."[19]

On the whole, however, middle-class race leaders were too closely tied to district employers to press the grievances of black workers aggressively and settled ultimately into a familiar role as their labor recruitment (or, in the new context, retention) agents. Brandishing its unique amalgam of race pride and industrial submission, the *Birmingham Reporter* warned black workers early on against "the tendency to be persuaded away by unscrupulous and unreliable white labor agents" to "districts unsafe and unsettled from a labor view point." Echoing one of the most common stereotypes about black workers as "a people who like to move about from place to place," editors expressed their bewilderment at the flight. "Why our people would be persuaded to leave this district is rather puzzling," they complained, urging blacks to "settle down" and "stay in Birmingham." The same commitment to ending the exodus of black labor out of the South was evident in the position adopted by authorities at Tuskegee Institute, with whom Birmingham's race leaders maintained close relations. Acting in cooperation with local planters, Tuskegee officials "decided to make a determined effort . . . to persuade Negro farmers to remain on the land instead of going to the cities."[20]

With few exceptions, African-American religious leaders openly opposed the exodus and came to the aid of district employers, admonishing black labor to "stay put." Delegates from the African Methodist Episcopal Church, sounding an alarm against the level of "unrest and dissatisfaction" evident among blacks during their 1916 convention in Birmingham, attributed the scale of the migration to frustration with racial prejudice, disfranchisement, and Jim Crow. Individual ministers boldly defied southern white opinion, rejecting popular explanations that ascribed the exodus to alleged racial deficiencies and insisting instead that it was the southern social order and not blacks' moral character that was in need of urgent reform. On the whole, however, the black church played a conservative role, one for which the script seemed at times to have been written by industrial employers. The *Reporter* noted in December 1916 that from the "pulpits . . . of many of our leading churches, the colored peoples' attention ha[d] been called to the raise of wages throughout the district." "It is hoped," the editors wrote, "that as the time goes on our people will settle down and become satisfied with conditions." Denouncing the "disgrace" of black emigration as a product of the race's "ignorance," Adams warned on another occasion that the exodus would "put a stain, a suspicion on Negro labor that has never been on it before." "How can we expect the decent consideration of influential honest white citizens," he asked, "when we so deport ourselves?" In the absence of a more aggressive challenge to white elites, the church—along with the black middle class generally—preached the familiar message of moral uplift. Proud of the fact that "nothing has been more influential in remedying this restlessness than the church," ministers admonished their flock to "lead temperate lives" and "rally to the church as the Ark of safety on this troubled sea."[21]

Coal camp preachers played an even more explicit role as propagandists on behalf of the major operators. R. W. Weatherly of the West End Baptist Church at TCI's Ishkooda camp issued a public warning to "the Colored Citizens of Industrial Coal Mines, Ore mines, and Other plants," which was printed in the black press and circulated throughout the district. Warning black miners against being lured away to the coal districts in Kentucky and Virginia, where conditions were "not favorable to the colored people," Weatherly advised them to remain in the Birmingham district, where, he contended, "conditions [are] better . . . for colored people than any other place in the Union." The same message was carried throughout the camps by the corps of black welfare workers that had been recruited since 1908. Robert W. Taylor, a Tuskegee graduate, the principal of the colored school at Sipsey, and a paid employee of the DeBardeleben Coal Company, addressed a meeting of black and white miners in September 1916, advising them against being "deceived by the oily tongues and glittering promises of well paid labor agents." Sharing the podium with the Sipsey supervisor Milton Fies, Taylor urged those "who like peace and security . . . while earning an honest dollar in the sweat of their brow" to remain in Alabama.[22]

The employers' concern for retaining their primary labor source presented black Alabamians with a unique opportunity to force white supremacy onto the defensive. Although prominent race leaders were not oblivious to these new possibilities, their strategic alliance with white elites and their faith in middle-class morality as a solution to the plight of ordinary blacks rendered them incapable of providing determined leadership. In the face of ever more desperate conditions for the mass of black Alabamians, race leaders remained confident that individual perseverance could overcome impediments to racial advance. Before war demand had begun to lift the national economy out of depression in early 1915, local black insurance agents complained that the unemployed were submitting fraudulent claims to help them through hard times. Oscar Adams denounced the "too pronounced disposition on the part of some of our people . . . when work is difficult to obtain and wages rather low, to take advantage of any little thing that will bring to the family a few coffers . . . each week." In the early stages of the exodus, Adams dismissed the notion that it was driven by actual hardship, blaming the unrest instead upon the fact that "not only the American Negroes, but the American people, have formed a habit of living beyond their means." Not the dull prospect of continued poverty and oppression, he argued on another occasion, but "the brilliant star of hope is shining today for the southern Negro and he should remain in the South to partake of its effulgent rays." The *Reporter* reprinted an editorial by Frank Evans that counseled black miners against leaving behind the "peaceful relations which exist between operators and miners in the Alabama coal fields" for the North, where "agitators have been busily at work with the result of idleness and disorder, lawlessness and tragedy."[23]

One feature of the campaign to prevent black out-migration revealed clearly the intimate relationship between leading coal operators and Birmingham's black middle class. Among the infamies attributed to the northern fields, the black press harped incessantly upon the charge that blacks emigrating to West Virginia and Kentucky had been subjected to racial abuse, implying that such treatment was foreign to the Birmingham district. On close examination it is clear that their complicity in publicizing such complaints derived from their enlistment in an elaborate propaganda effort orchestrated by district coal operators. Praising local employers and the ACOA in particular for "getting along with its labor admirable well," the *Reporter* juxtaposed the placid labor situation at Birmingham with regular coverage of racial clashes in the North. "There was never a time in the history of coal mining in Alabama," it contended in November 1916, "when relations between miners and the operators were more friendly, cordial, and remunerative than just now." In contrast, the editors cited recent rioting between blacks and whites at Wyandotte, Michigan, attributing this friction to the *absence* of Jim Crow: "the fact that up to now no provision is made for adjusting the relations between the two races works undue hardships on the southern black man." "Alabama," they concluded, where the color line was more clearly etched, "is the best place for colored labor."[24]

Complaints about mistreatment in the North were not without substance, of course. In their efforts to direct the flow of emigration, northern operators dispatched labor agents, both white and black, throughout the Birmingham district. Promising higher wages and superior conditions and offering to pay transportation costs out of the district, these emissaries met with an enthusiastic response from experienced Alabama miners squeezed by the new influx of cheap labor. Within months of the first wave of out-migration, however, reports of mistreatment and broken promises began to circulate throughout the district. The U.S. attorney for northern Alabama, Robert N. Bell, reported in October 1916 that "the inducement being offered" Alabama miners by West Virginia operators was "fraudulent in nature." "They are told that they will get 42 . . . cents for loading a car of coal at these coal mines and, when they reach the place, they find that the car has twice as much capacity [as] the car used in this district." A number of black and white miners complained of being held in peonage in northern coal camps, and black miners in particular expressed their disappointment that the North had not lived up to their hopes. A black miner returning from Hutchinson, West Virginia, was disappointed by the lack of "churches or schools for colored people." Another who had spent time in Kentucky resented having to "mix" with "a low class of foreign white people."[25]

While conditions in the northern coal districts undoubtedly fell short of the generous regime depicted in the labor agents' recruiting pitch, the charges against northern operators are rendered suspect by the prominence of the ACOA and its black middle-class allies in attempts to terrify the district workforce into staying put. Observers noted several years later that industrial employers were consciously engaged in an attempt to "stimulate a tide of return" of black labor to the South, and the ACOA's efforts during 1916 and 1917 constitute one of the earliest examples of this. Though Birmingham operators resented the intrusion of northern-based labor agents into the district and demanded vigorous enforcement of laws designed to insulate them from such competition, they had never shied away from resorting to identical recruitment methods themselves. A TCI manager informed Urban League representatives in 1916 that "his company had spent fifteen thousand dollars importing Negroes, principally from Kentucky, Tennessee, West Virginia, and Virginia." The purpose of this operator-induced, reverse migration (in which the ACOA played a direct role) seems to have been "to have these men come back and tell about conditions in the [northern] mines so as to check further movement North." It was for their value in stemming the exodus, therefore, "rather than the scarcity of labor," that operators encouraged native miners to return.[26]

The partnership between the ACOA and the black middle class was crucial in disseminating the message about poor conditions in the North. The crowning event of the Sipsey meeting at which the black school principal R. W. Taylor and the mine boss Milton Fies shared a podium before a racially mixed group of miners was the testimony from a group of "white and colored miners who told of their bitter experiences in the coal mines of Virginia and West Virginia." The Urban League reported rumors that "Negro newspapers have been subsidized to print stories about large numbers returning and suffering in the North," a suspicion that seemed confirmed by coverage in the *Reporter*. The testimony cited above, from black miners who had recently returned from Kentucky and West Virginia, appeared in that paper

along with a half-dozen other interviews under the heading, "MINERS RE-
TURN TO ALABAMA AND TELL SOME DISTRESSING STORIES; WILL
STAY AT HOME." Accompanying their accounts was a notice that local coal
companies were about to raise wages.[27]

The cynicism with which black miners viewed these efforts can be seen not
only in the failure of the operators' campaign to halt an exodus out of the
district but in the distinct absence of loyalty or gratitude among those taking advantage of employer sponsorship for the return trip. Federal officials
investigating the return migration noted with some irritation that they had
finally succeeded in obtaining testimony from returned TCI employees "after much persistency and many promises" from company officials. TCI management delayed forwarding a final transcript of this testimony "day after
day," investigators claimed, "[because] they wanted to cut some things out
of it, which evidently was done." Urban League field workers who interviewed
returning miners found that, contrary to the reports being circulated by the
ACOA, "none of those who had returned gave disparaging accounts of the
North nor were they dissatisfied nor did they intend to stay in the South."
Struck by the fact that "despite the large numbers who were reported to have
returned, no appreciable difference could be detected in the labor supply,"
investigators found not only that these numbers were inflated but that in
many cases black miners "were simply saving transportation one way by allowing the [Alabama] firms to bring them back . . . and when they returned
they . . . took their families and left to stay."[28]

In the eyes of ordinary blacks, the severity of the crisis that had spurred
mass migration rendered moot and even farcical the middle class's prescription of hard work and sobriety as a remedy for their plight. The collapse of
race leaders' authority was accelerated by a number of developments peculiar to the Birmingham district. The failure of the black-owned Alabama
Penny Savings Bank, long held up by accommodationists as a model for race
progress, and the "almost destruction of a number of [black] fraternal organizations" by a series of financial scandals dealt a body blow to race leaders' self-assurance and injected a sense of deep malaise into welfare work. One
keen analyst attributed the bank's collapse to its being "caught in the back-
wash of steel and mining troubles of the time," and its acting president after
Pettiford's death in 1914 explained that "customers who were poor and many
of them out of employment . . . could not pay their interest much less the
principal." One black woman described the failure years later as "the most
tragic thing. There were all the old people out there crying [who] wanted their
money." The sense of crisis found its way into the pages of the *Reporter*.
"What is the matter?" Adams wondered. "Leaders are dying, banks are fail-

ing, institutions of pride, church organizations, fraternal organizations, are splitting . . . confidence in the Negro to handle large sums of money is shaken, while the gambler, the alarmist and the unholy agitator go on with seeming prosperity."[29]

The demoralization evident among Birmingham's most prominent race leaders reflected the massive transformation that the upheaval had inaugurated within the district's black community. Although the subjection of all southern blacks to pervasive racial hostility heightened the appeal of racial solidarity and blunted intraracial class antagonisms, the divisions that had surfaced during periods of working-class mobilization confirmed that this appeal had limits of its own. Migration shook loose the bonds between the black establishment and black workers, jarring the confidence of race leaders who feared that they had lost their grip. R. W. Taylor reported worriedly from Sipsey that "the Negro labor of the Alabama mining district is an incoherent mass, united in nothing save the conviction that Capital has the 'cards stacked and marked against it.'" The contrast between the poverty into which the lives of most black southerners had deteriorated and the relatively comfortable material position of the black middle class impressed itself upon Urban League officials, who found that while some Birmingham blacks were "prospering and have been fairly secure . . . this applies more particularly to the established Negro property owners and entrepreneurs" than the "laboring classes."[30]

Observers of the migration out of the South universally noted the spontaneity of the movement and its lack of a coherent leadership. The Reverend J. Bolivar Davis, sent south by his superiors in South Carolina for the purpose of "preach[ing] to the colored people," found it impossible to "get up a meeting." "Everybody is restless here in Alabama," he reported. "The Northern fever is raging down this way." Another minister, reporting from Georgia, found that "the colored race, known as the race which is led, has broken away from its leaders." Blacks heading north from Alabama, the *Reporter* warned defensively, "are practically without a leader." They were departing against "the advice of many of the best men of our race who are able to give them wise and wholesome instruction." Black workers had "broken off from consulting the men of the colored race, and . . . learned to scorn the advice of men of the white race."[31]

The most telling evidence of black workers' rift with established race leadership was their collective rejection of elite pleas to stay put. Ordinary blacks divined in the concurrence of natural and economic disasters not tragedy but the hand of an avenging God. One observer noted that blacks appeared to "feel that the boll weevils and floods are God's agencies of deliverance for him, from a system of agricultural bondage, in some respects and in some sections, more despicable than actual slavery." Another wrote of a "general belief that God had cursed the [South], for nothing would grow." Many of those heading north shared a belief that "God is in this movement" and "spoke of leaving Birmingham for 'God's country.'" For many, the black middle class's complicity in efforts to check their exit at a time when the coal companies were attempting to do so through open coercion offered final proof that race leaders were hopelessly compromised by their ties to Birmingham's industrial interests. The bitter parting shot of one migrant heading North illustrates how closely these two groups had become linked in the view of black workers. "The Negro papers which you subsidize and Negro leaders whom you pay, cannot hold [us]," he vowed. "Two are [*sic*] three years ago you promised us schools; you have not given them to us. The only thing you have offered us is an old Jail for our children." At least one member of the clergy concurred. Birmingham race leaders, he concluded, "can no more keep Negroes here than they can fly to heaven backwards."[32]

On the eve of the United States' entry into the war, therefore, the system of racial paternalism developed by operators after 1908 had begun to collapse under the strain of black disaffection. Increased war demand presented Alabama blacks and whites for the first time with prospects of industrial employment beyond the mines and mills of Birmingham, and unprecedented competition for the native labor supply had begun to strip district operators of their hegemony, creating new openings for working-class organization. The nation's formal entry into the war in April 1917 raised the stakes even further: the siphoning off of native labor in even larger numbers, for employment in the North or service in the military, exacerbated the growing labor shortage and set operators and union miners on a collision course. Inspired by the rhetoric of Woodrow Wilson's "war for democracy," miners would attempt to extend national war aims to include the triumph of industrial democracy at home. Meanwhile their employers—invoking another brand of patriotism—would seek to rein in renewed militancy by adapting the repressive measures enacted against wartime dissidents for use in the coalfields. Both sides, therefore, scrambled to take advantage of the new possibilities opened up by the war, injecting a volatility into coalfield labor relations that would erupt eventually in explosive confrontation.

Even as it rescued operators from the sloth of depression, economic revival made it more urgent that they resolve the problems associated with labor supply. U.S. entry into the war "immediately changed the balance of forces in the Birmingham district," Joseph McCartin writes. "War orders removed the impediments that had previously constrained the growth of steel production in the South." Increased demand made itself felt almost immediately in the coal industry. From the "dark period of depression," about which the ACOA had complained in early 1914, "the coal fields ran to capacity as demand . . . rocketed after 1916." Urban League field workers found that "mines that have not been in operation for twenty-five and thirty years are opening," resulting in a situation where "jobs sought workers—not the reverse."[33]

The UMW's energetic campaign to revive union organization in the Birmingham district ruled out for mine operators the luxury of approaching their new predicament as if it were simply a matter of more effective labor recruitment. In a context where their traditional supply was bypassing district mines and heading north and where those who decided to remain harbored raised expectations about their treatment and compensation, district operators could not realize the fruits of industrial upturn without making it reasonably attractive for miners to stay put. In spite of their strong aversion to change, operators were forced to advance wages in step with those on offer in the North. And with the framework of union organization in place from June 1917 onward, whatever concessions they did grant were made with the threat of industrial action looming over the district. Just as the necessity of maintaining a low-wage regime had been imposed on them by competition with the northern coalfields, during the war Alabama operators were forced to concede wage increases to counter the allure of northern industry and the gathering threat posed by a reinvigorated UMW.

The war boom affected Alabama's coal industry unevenly, however. The granting of wage increases by the leading captive mining companies imposed a financial burden upon smaller, commercial operators. In a confidential letter to the TCI president George Crawford in August 1916, the Alabama Senator John Bankhead, acting as an intermediary for independent operators, expressed their dissatisfaction that those "companies producing iron and steel have received all the benefits accruing from the increased demand," while the "commercial coal operators have not shared in the war prosperity." Increased profits among the large captive operators had led naturally to "some [wage] increases to mine laborers" with "further increases suggested," while smaller operators who "had not increased their earnings, but [only] their costs" found it "impracticable . . . to meet the increased wages." The resulting tension in the operators' ranks threatened to leave them divided in the face of the developing UMW challenge. "Much restlessness had developed [among miners]," Bankhead warned Crawford, "and . . . it is only a question of time when the trouble will culminate in a serious effort to organize . . . followed with a strike." War prosperity therefore generated fractures in the operators' ranks even as it pulled the district out of a slump: mine owners were sharply divided over the wisdom of material incentives as a prudent means of holding their labor force.[34]

Even among leading operators, their highly visible efforts to slow the exodus through wage raises and a coordinated public relations campaign never precluded the use of coercive measures. Nor were they prepared, after 1916, to let the UMW roll into the district without mounting a determined fight. They entered the war years having developed a substantial security apparatus in the period since 1908, an apparatus founded with the aim of insulating the mine workforce from "outside interference," whether in the guise of out-of-state labor agents or the contagion of trade unionism. Not surprisingly, the operators responded to the first wave of migration with traditional, heavy-handed methods of labor control. Attempts to hamper the mobility of miners, like the incident involving black miners at the railroad depot in June 1917, became more routine. Mine bosses intensified their vigilance against union organizers and expressed their fierce objection to the intrusion of out-of-state labor agents, calling upon state authorities for stricter regulation of such activity. Citing an "acute labor shortage," twenty-five "prominent iron and steel manufacturers" prevailed upon the Birmingham city commission to pass a strict new vagrancy law. Livid over the fact that wartime wages permitted men "to work two or three days a week, and still support themselves," industrialists called for ridding the streets of those "guilty of wandering or strolling about, or remaining in idleness during any working day in any calendar week."[35]

With the declaration of war in April 1917, area industrialists sought to embellish such measures with patriotic ardor, hailing them as vital to the nation's mobilization effort. Though many worried that federal intervention in the labor market would have debilitating long-term consequences for the region, employers throughout the South hoped that federal restrictions on labor mobility would bolster their own efforts to stem the exodus. Their optimism was rewarded when the U.S. army provost marshal, Gen. Enoch E. Crowder, issued a "work-or-fight" order in May 1918, stipulating that all able-bodied persons of draft age must be engaged in necessary employment. The historian Philip Foner has enumerated the obvious advantages of such legislation for southern employers. "Statutes of this type," he wrote, "provided local officials with a convenient tool to prevent strikes, mass emigrations, or any other action that they could charge hindered war production." Southern industrialists had previously called upon the federal government to play a more active role in staunching the hemorrhage of their labor supply. Government officials expressed frustration, however, at being unable "to devise any way by which the migration can be controlled without at the same time interfering with the natural right of workers to move from place to place at their own discretion." The military's new directive provided a framework satisfactory to employers. After conducting a tour of the country, the assistant secretary of labor Louis F. Post announced that he had "found that the Work-or-Fight Order was being used for peonage purposes and that employers were conscripting labor for private use rather than for service to the war effort. This was especially true, he reported, in states where the majority of workers were black."[36]

Federal officials attempted to counter the "wild rumors circulated among colored people . . . that the new work-or-fight order . . . meant that men would be drafted for labor and put under conditions amounting to peonage," but with little effect, particularly since that is precisely what transpired in many parts of the South. Reports of peonage were "exactly coterminous with that portion of the territory of the U.S. in which the institution of chattel slavery formerly existed," NAACP officials asserted, finding moreover that "from the usual condition of the great mass of [laborers] where these laws are enforced to peonage is but a step. It is not unusual to find that . . . large numbers of laborers are restrained of their liberty in quarters and in stockades, guarded by men who carry guns and deadly weapons." Though government officials insisted that the work-or-fight order was as "different from peonage as self-dedication for national purpose is," NAACP officials came closer to the mark when they warned that "no one can tell what extremes Southern white employers will go as the exodus continues, to practically revert to slavery in getting labor."[37]

The chief benefit of the government's work-or-fight order to southern employers was that it came at a time when they were "looking for some pretext to handle Negro labor and at the same time avoid [the] appearance of oppression." Crowder's directive emboldened state and municipal legislatures throughout the South to pass local ordinances restricting the mobility of (mostly black) labor and bestowed an air of legitimacy and even patriotism upon transparently repressive methods. After an eight-week tour of the region, Walter White of the NAACP emerged "convinced . . . that the South is totally blind so far as its Negro labor is concerned." Racial prejudice blinded the white South from seeing the "error of harsh methods in handling Negro labor," he thought. The region seemed oblivious "to the changes in the status of all labor and in its case particularly of Negro labor . . . brought about by the war. Ante-bellum and ante-war methods yet are considered proper."[38]

An atmosphere of increasingly strident militarism and nationalistic fervor provided the context in which such methods could be applied without unsettling the conscience of either the white South or federal officials. ACOA officials reported in 1918 that their members had "actively taken part and rendered all assistance possible in all patriotic movements and auxiliary war work," taking "the lead . . . in every county in the coal fields." A series of bulletins distributed among employees of the DeBardeleben Coal Company captures the tone of the operators' campaign to identify increased productivity with patriotic duty. Bulletin number 2 asked miners: "Are YOU producing all YOU can? / Are YOU making any sacrifice to produce more? / Are YOU going early and staying late? / Do YOU load one car more even if YOU have loaded all you want?" and ended by challenging each employee to "BE A MAN, NOT A QUITTER." The effectiveness of such appeals— backed by mild coercion—can be gleaned from a report on the situation at Margaret mines: "Instead of the men going into the mine in the morning and coming out at 2 in the afternoon," a *Reporter* correspondent observed, "they now stay until 5 o'clock. They are made to feel that there is nothing so important now as coal."[39]

Bulletin number 10 featured a poem, "The Miners' Resolve," which recounts the story of two miners who attempt to enlist in the army. One of them, rejected because he was missing two fingers, is left behind to work in the mines while his partner, Jim, goes off to fight in Europe. The story was obviously intended to impress upon miners the vital role that they could play in the war effort: "'So Jim has gone, good scout, and left the hole / Them fingers, well it kinder makes me blue / But shells, they say, depend on lots of coal / And while Jim fights for both, I'll dig for two." "Uncle Charlie" De- Bardeleben, owner of the Alabama Fuel and Iron Company, employed a more intimate form of persuasion to increase output at his Margaret mines. De- Bardeleben regularly made a "man-to-man canvass [of his employees] . . . and when he finds [one] who is working but two days, he asks him why he does not work six—not for the money . . . but because the government needs the coal." DeBardeleben thus made productivity "a patriotic matter, and asks the man why he is not doing his duty."[40]

Relatively benign attempts at coaxing increased production out of the mine workforce leaned upon more forceful means of compulsion. In the industrial setting of the Birmingham district, the rampant jingoism unleashed across the country after U.S. entry into the war, directed elsewhere against naturalized German-Americans and left-wing labor activists, was harnessed to the production goals of district employers. Within months after passage of the city's strict vagrancy law, employers expressed their satisfaction that the combination of new local legislation and the federal work-or-fight order had materially augmented the labor supply. "Pool rooms have been deserted [and] few loiterers are seen in the streets during work hours," city officials reported. "The negro section, where once we found scores loafing, now look[s] like a cemetery, and Birmingham is carrying out to the letter the 'work or fight' order." In nearby Bessemer, authorities erected a "Slacker's Cage" in the center of town to expose those charged with undermining the war effort to public ridicule. Its first occupant, a white mine carpenter named Hayes Brown, found himself caged after TCI foremen fingered him for "disloyal utterances against the Government" and "nonsupport of the war burden." Brown became so despondent after being subjected to the taunts of some five thousand passersby that he reportedly "attempted to place his head in the loop of a rope with which he was bound to strangle himself" before being taken into protective custody by local police and brought to the city jail. Brown's only crime seems to have been his unwillingness to consent to a wage deduction for the purchase of Liberty Bonds.[41]

The full import of the employers' determination to turn the war hysteria to their own advantage became clear during a district-wide steel strike in 1918. Belatedly cognizant of the debilitating effect that racial divisions had exacted on previous organizing attempts, white metal trades' unionists called upon the International Mine, Mill, and Smelter Workers' Union to help organize unskilled blacks in the steel industry. As the union drive gathered momentum, district employers unleashed a campaign of violence and intimidation, directed disproportionately at black workers. In April 1918 vigilantes dynamited a house occupied by the black Mine Mill organizer William Hale; two months later he was beaten and tarred and feathered for "preaching the industrial equality of the Negro," a doctrine that his attackers believed would "cause labor unrest, which was unpatriotic."[42]

The violence was accompanied by renewed exhortations from employer-sponsored black welfare workers, who urged blacks to steer clear of the organizing effort. The black TCI labor agent Melvin Chisum wrote friends at Tuskegee that he had been "engaged for weeks in careful preparation for any untoward situation which may arise" due to the strike. P. Colfax Rameau's *Workmen's Chronicle,* published out of Chisum's office at TCI headquarters, urged strict enforcement of work-or-fight laws to round up the "human parasites . . . and Industrial slackers" of both races who were "hiding behind this damnable unionism operated by Pro-German sentiment." The local NAACP official Charles McPherson reported that leading employers were "endeavor[ing] to organize among the Negroes a union of their own" but noted, significantly, that such attempts met with "some opposition" among black steelworkers.[43]

Despite the employers' efforts to forestall interracial cooperation, white union officials reported in April 1918 that they had "at last succeeded in making some progress toward organizing the negro and common laborers." They predicted that "the situation will become acute in the near future" owing to the determination of district employers to "prevent the organization of this class of labor." Into this volatile situation stepped the recently organized National War Labor Board (NWLB), offering steelworkers a "channel through which [they] could bypass the local power structure dominated by their employers."[44] It was this aspect of the NWLB's intervention in Birmingham that district employers regarded as outrageous. When the board set a date for hearing both sides in the Birmingham dispute, industrialists considered their action a "profound federal intrusion" into the district's "private affairs and traditions."[45]

The resurrection of the Ku Klux Klan in Birmingham was, according to McCartin, a direct response to the NWLB intervention. One hundred fifty robed Klansmen marched on horseback through the city's streets on May 6, 1918, bearing a message "unmistakably directed at striking steelworkers, who marchers termed 'idlers and disloyalists.'" The Klan attempted to fuse its historic role as the vigilant guardians of white supremacy with the new xenophobic impulse sweeping the nation, passing out leaflets declaring that they were "on the lookout for alien enemies, for the disloyal and for the fellow who is seeking to begin a strike," and warning "slackers" to "work or fight."[46]

The important link between the employers' hostility to labor organization, their fear of black militancy, and the resurrection of organized, extralegal violence can also be seen in the rise of a paramilitary organization known as the "Vigilantes" during the thick of the agitation in steel. Concerned over the growing appeal of unionism among black steelworkers, prominent Birmingham citizens—with the apparent consent of high-ranking federal officials in the district—organized themselves to nip the organizing drive in the bud. The NWLB examiner Raymond Swing observed that the Vigilantes did not try "to interfere with organized white men, but have been used almost exclusively to frighten the negro employees." McCartin recounts their exploits:

> At several meetings of black workers the companies sent gunmen in cars who parked with headlights shining on meeting house doors taking the names of those who entered. Armed gunmen broke up a meeting of black steelworkers at Tuxedo. Three separate organizing meetings in Ensley Machinists' Hall were broken up by armed gunmen preventing blacks from entering. . . . Similar events occurred in Mason City. At Bessemer, unions succeeded in getting municipal authorities to pass an ordinance forbidding interference with labor meetings, but they could find no one to enforce the law. At Avondale, a black woman who owned a meeting hall received a bomb threat that warned her not to allow union meetings in her establishment.[47]

The Vigilantes' selective targeting of black steelworkers reflected Birmingham employers' historic concern that the contagion of trade unionism not be allowed to infect their principal source of cheap labor. In every major confrontation with organized miners, district coal operators had pinned their hopes for defeating unionism upon their ability to seal off the mass of black miners from the radical ferment sweeping the coalfields. Significantly, the 1918 events in steel marked the first time that white steelworkers reached out to organize unskilled blacks, a tactical reorientation that provoked a panicked reaction from their employers.

Though they occasionally crossed swords with federal officials during this period, the employers' ability to link labor militancy with a perceived threat to white supremacy and to conjure in the public psyche a palpable fear of black insurrection owed much to the actions of the federal government itself. The white South's fear of racial insubordination found a striking parallel in the public disposition of prominent federal officials. By 1918 no less than five federal investigative agencies were "engaged in a detailed secret scrutiny of black American opinion and leadership." The Department of Justice charged publicly that "German propagandists" and labor radicals were aggressively seeking to take advantage of disaffection among African Americans, and in 1919 the NAACP was compelled to denounce the "pernicious campaign . . . being carried out to [suggest that] the IWW were responsible for . . . discontent and dissatisfaction among the Negroes." A report to the director of military intelligence warned that "radical propaganda ha[d] made noteworthy headway among the colored people in this country." "Beyond a doubt," investigators cautioned, "there is a new negro to be reckoned with in our political and social life."[48]

Area employers and federal authorities stationed in Birmingham were determined to prevent this "new Negro" from upsetting the district's status quo. Occupying one of the central points of departure for blacks exiting the South, Birmingham's black citizens found themselves subject to intense scrutiny by federal authorities, whose investigations uncovered scant evidence of organized discontent. Justice Department officials imported a black undercover agent into the district after an altercation aboard a Birmingham streetcar left a white policeman and a black passenger dead, convinced that the rise in racial tensions "must be due to some outside influence and agitation." After some fairly laughable antics ("have loitered around public places in order to pick up radical agitators, but have found none"), the special agent, W. L. Hawkins, succeeded in locating a "large number of guns and ammunition stored within ten minutes walk" of the city's Negro Quarter but concluded that the weapons were intended for defensive purposes and otherwise found "very little evidence of coordinated plans."[49]

In the volatile atmosphere of wartime Birmingham, however, the lack of hard evidence of subversion scarcely alleviated white fears that something was afoot. "Despite statements of loyalty from many leading Alabama Negroes," Richard Eaves has written, "the press continued to report rumors of possible trouble from them in regard to the war." A delegation representing twenty black organizations marched to Birmingham City Hall at the outbreak of the war to pledge their "total cooperation in the fight against Germany." Oscar Adams and a host of lesser known race leaders were prominently involved in recruiting blacks for military service and selling war bonds throughout the black community; some went further, offering federal authorities their services as informants on black antiwar agitation. Nevertheless, the media remained focused upon real or imagined examples of black disloyalty. The *Birmingham News* reported in April 1917 the case of two African Americans reported to have "urg[ed] fellow Negroes to rise up and take over the country after the white men went to war." Their "bullet-riddled bodies were found in a ditch" a short time later. The *New York Times* announced almost simultaneously the arrest of two white men at a mining camp at Corona, "accused of having made speeches tending to incite the blacks against the American government," along with the arrest of a black man who allegedly "made speeches to fellow-members of his race, in which he urged them to denounce this Government and turn their efforts on behalf of Germany."[50]

The conflation of racial and industrial militancy in the eyes of Birmingham employers renders it difficult to determine whether authorities were dealing with authentic cases of sedition or merely taking advantage of wartime indulgence in vigilantism to clamp down on workers' rights. The controversy that arose during an ACOA-sponsored speaking tour in the fall of 1917 illustrates this point. Presenting herself disingenuously as a representative of "Mr. [Herbert] Hoover, the Food Administration, and the government,"[51] Mrs. G. H. Mathis addressed a number of mass meetings across the district, urging miners to aid the war effort by raising productivity ("Work early and work late and let every pound of coal count," she urged miners) and "hunting down" the German spies in their midst. "We must go lion hunting," Mathis beseeched a meeting of a thousand miners in Walker County. "We must search them out here at Empire, in Birmingham, in every nook and cranny of America." According to the *Age-Herald,* Mathis paid a "glowing tribute" to the "assembled miners blackened with grime," but in private correspondence with Labor Department officials she claimed to have detected a "very considerable German influence at work in the mining districts of Alabama." Significantly, she warned that the "propaganda" was "using the miners' Union as a channel for reaching the mining class" and that it "appear[ed] to be getting a hold on the negroes," who were "acting with a show of threat towards the white people."[52]

As an example of pro-German sentiment in the coalfields, Mathis cited the report of a draft board examiner who told her that "of four hundred men that had appeared before the Board, three hundred and ninety-nine wanted to be excused." Over 50 percent of the population in the mining district, she estimated, did not seem to "understand what the war is about and why we are in it." Mathis's observations were confirmed in a communication received by Governor Charles Henderson after he commenced the organization of a state Home Guard in the summer of 1917. A Gadsden attorney wrote the governor, warning him to "avoid the organization of elements of our population and their equipment by the Government who may stand in opposition to [its] aims." A resident of Pyriton wrote seeking Henderson's advice regarding the organization of the guards in that vicinity. Among those working in the area's mines and turpentine camps, he warned, were "numerous pro Germans" who were reportedly "organizing themselves against the draft act."[53]

While their heightened surveillance of miners' activity during the war years was carried out under the pretext of defending national security, it is clear that the operators' primary concern from 1916 onward was to prevent the UMW from regaining a foothold in the district. The centerpiece in their efforts to maintain the open shop was the blacklist: from the beginning of District 20's reorganization in early 1916, ACOA members conducted an ongoing purge of union sympathizers throughout the coalfields. In April 1916 the labor press reported the "discharge of several good miners" after a superintendent from Gulf States Steel presented the operator of a small independent mine with a list of men he believed to be tied to the UMW organizing effort. The District 20 president J. R. Kennamer wired the Labor Department in August 1917 after TCI discharged over five hundred suspected union sympathizers. That company's much-vaunted record of racial benevolence apparently did not extend to those who supported the union: of 345 miners blacklisted in the first seven months of the year 129 were black. An internal company document obtained by the UMW revealed that African Americans made up nearly half of the thirty miners discharged for "insubordination" and "agitation" by TCI over a three-week period in July 1917. And the operators supplemented their purge of suspected UMW sympathizers with close scrutiny of those left behind in the camps: a black TCI employee from Belle Sumter, for example, who "hat[ed] to see [his] race don [sic] so bad," wrote to the Labor Department complaining that management was forcing black miners to sign "affidavits" pledging to work through any industrial dispute that might arise."[54]

To the extent that measures like the work-or-fight order or the more general wartime circumscription of workers' rights afforded them greater latitude in combating trade unionism, the coal operators welcomed federal intervention. Though consistently hostile to federal agencies such as the NWLB or the U.S. Employment Bureau, which they perceived as threatening their unilateral control,[55] and never completely reconciled to the new state of affairs ushered in by the war, Birmingham's industrial employers developed a discriminating approach to government involvement. The outcome of the steelworkers' strike, in which a deadlocked NWLB ultimately backed off from intervention on behalf of the union,[56] and the willingness of individual government officials to indulge Birmingham employers in adapting patriotism to production requirements[57] convinced some that a limited federal role could be tolerated and perhaps even deliver real advantages. A few reform-minded agents aside, it was not the intention of federal officials to upset the southern status quo: federal bodies proved quite amenable to the interests of capital,[58] and the ACOA apparently reasoned that a more cooperative approach would pay greater dividends. In 1918 four Birmingham coal companies lent high-ranking members to serve on the staff of the U.S. Fuel Administration in Washington and paid "for several months, the entire expense" of its representative in the district. Alarmed, perhaps, by the steelworkers' bold (and nearly successful) attempt to wrest sanction for their organizing efforts from the NWLB, the operators came to understand that their interests would best be served not by ignoring federal intervention but by using their considerable influence to render it amenable to their own plans for the district.[59]

Ultimately, however, industrial Birmingham's efforts to shape federal labor policy could not deliver them from the essential threat to their hegemony. While the probusiness disposition of individual federal officials or even entire agencies might soften its blow, the war's dramatic undermining of the South's closed labor market could not be averted. The key to the employers' success in maintaining a low-wage regime before the war had been the lack of options available to black and white laborers elsewhere. Hard-won subsistence on the dirt farms of northern Alabama; monotonous, oppressive, and frequently unpaid labor on the cotton plantations of the Black Belt; or slightly better conditions in the industrial complex growing up around Birmingham: these were the bleak alternatives from which ordinary Alabamians had to choose, and it is not surprising that many opted for paid employment in the mines and mills.

Wartime demand had loosened if not completely broken Birmingham industry's stranglehold over the region's labor supply—a fact that no amount of maneuvering could alter. The employers' frustration derived less from disappointment with specific actions of government officials than with the general trend that their intervention portended. District employers continued to regard northern industry as the chief culprit in drawing away their labor supply but inevitably began to detect the federal government's hand in the tampering. Wage hikes that might otherwise have been considered an inevitable result of increased war demand were attributed to the Wilson administration's allegedly cozy relationship with organized labor. Locally, the *Labor Advocate* reported that Birmingham's "powers that be" rejected government plans to build an armor plate mill because it would "upset the labor conditions in the district." Sloss-Sheffield executives lamented the impact of federal policy on local wages. "Never . . . in the history of your company," the president of the company, J. C. Maben, informed stockholders in March 1919, had there been "a more unsatisfactory condition of labor as to its scarcity, its wage, and its efficiency." Over seven hundred Sloss employees had enlisted in the military, he recalled, while construction work at local government munitions plants had "raised the scale of wages for common labor . . . one hundred percent." Judgments by the Railroad Administration and the NWLB had raised wages even further, producing "a migratory condition of labor which resulted in demoralization, indifference and inefficiency."[60]

From a situation where, following their triumph in September 1908, district operators had more or less controlled the movement of miners from camp to camp, by 1918 they were no longer in a position either to prevent men from playing local operators against one another or abandoning the district entirely. Nor did recent arrivals' relief at having escaped the cotton economy reconcile them to the conditions that the operators hoped to impose. John Garner, a new arrival, reasoned that while the boll weevil "freed us away from the landlords . . . we still wasn't free because we became wage slaves after we got . . . here."[61] Miners throughout the district understood from the outbreak of war in Europe the precariousness of the operators' position and seemed eager to take advantage of the change in circumstances. The rapid growth of the UMW during the spring of 1917 attests to an almost universal resolve among them that the catalog of injustices they had borne through the long years of operator ascendancy would be promptly remedied now that they had regained the upper hand.

Just as the operators attempted to associate their maintenance of labor discipline with the national patriotic crusade, Birmingham trade unionists—

including the UMW—sought sanction for their own struggles by publicly identifying with President Wilson's "war for democracy." Several months into the UMW reorganization, editors at the *Labor Advocate* celebrated the fact that Wilson's "declaration that the world shall be made safe for democracy has traveled clean around the globe and . . . finally reached the coal camps of Alabama." "If autocracy must go," they reasoned, "let us set the world an example by first cleaning house here at home." The ambiguity in Wilson's war rhetoric allowed Birmingham's industrial workers to turn the tables on their employers; they derided the operators' determination to deny union representation as evidence of industry's elevation of profits over national interests. In practice, labor rejected the business community's call for a suspension of industrial action during the war, countering with the argument that district employers had it in their power to prevent any disruptions. Individual unionists seemed not merely willing but eager to take advantage of their employers' new difficulty. "If we must die fighting let our face be toward our foe," wrote one, "and he is in our midst as much as he is confronting our boys in the trenches in France."[62]

Organized labor's attempt to match the employers' patriotism exacted a heavy toll on its esprit de corps, however. Prior to American entry into the war, Birmingham trade unionists, along with organized labor elsewhere across the United States, generally opposed the war as a "European squabble" from which the nation should steer clear. Throughout 1914 and 1915 Birmingham's labor press remained adamantly opposed to American military involvement in the conflict. Antiwar correspondence from pacifists and socialists appeared regularly in the pages of the *Labor Advocate,* and as late as 1916 its editors labeled the war a "new campaign of world conquest" on the part of "Big Business."

The seemingly wide-ranging debate in the labor press concealed the relative weakness of the Left in the local labor movement, however, and immediately upon formal U.S. entry into the war the conservative, lily-white majority in the Birmingham Trades Council forcefully asserted its dominance. When in March 1917 the Socialist Party member James Doyle of the Machinists' Union managed to pass a resolution in the Trades Council that "characterized the war in Europe as a contest for profits and urged that the rich do the fighting," conservatives denounced the measure as a "very smooth piece of propaganda work by a red-card socialist" and campaigned to have it rescinded. They published an advertisement in the following issue of the *Labor Advocate,* calling upon "defenders of the flag" who wished to "be relieved of the stigma placed upon [trade unionists] by the action of the Trades Council" to "organize a battalion of Union Men for service in [the military]."[63]

Although the industrial militancy gathering steam could not be completely derailed by labor's internal weaknesses, the silencing of labor's left wing and the capitulation of its majority to militarism carried important ramifications for the shape of the struggles looming over the district. The most obvious consequence of labor's shift to the right was its inability to maintain an independent profile in an atmosphere of increasing jingoism and reaction. The *Labor Advocate*'s prowar editorials became virtually indistinguishable from those appearing in the business press, and labor conservatives' endorsement of vigilantism against war dissenters rendered their protests against antilabor violence feeble.

The *Advocate*'s conspicuous endorsement of the term "slacker" to vilify those who had endorsed the antiwar resolution suggests that militarism had made deep inroads into organized labor. Likewise, the ambivalence with which Birmingham labor greeted the resurrection of the Ku Klux Klan and the employer-dominated Vigilantes after 1918 (particularly given their direct role in antiunion violence) illustrates how its embrace of militarism, leavened with white supremacy, had undermined labor's potential for mounting a unified resistance. After the Vigilantes' vicious attack on union organizers in June 1918, the labor-affiliated *Weekly Call* could barely muster a forthright condemnation of the affair. Editors expressed their shock that an organization that "numbers in its ranks some of the most loyal and just citizens . . . and has semi-official relations with the Government in ferreting out and ridding this section of German spies and American yellow dogs" would be involved in attacks on labor organizers. One doubts whether their suggestion that the Vigilantes continue to "make Birmingham hot for slackers and disloyalists and loafers" while keeping their "hands off workingmen who offended some profiteering corporation" appeared either feasible or appropriate to members of an organization that equated industrial militancy with treason.[64]

The militarist drift of organized labor had serious implications for one other critical aspect of its policy in the Birmingham district. Though in numerical terms they had never represented more than a small minority in the local labor movement, socialists and radicals had played an important role in legitimizing interracialism and posing an alternative to the lily-white modus operandi among white skilled tradesmen. Their marginalization tipped the scales further against the crystallization of a racially egalitarian current in organized labor. The practical necessity of organizing unskilled blacks in the district steel mills rendered out-and-out exclusion impracticable, and the resurrection of the UMW after 1917—in coalfields where African Americans now represented nearly two-thirds of the mine workforce—

offered an important counterweight to the conservative craft unionism embodied in the outlook of the Trades Council. Still, Birmingham's broader white working class, ensconced in a society more disposed to the ascent of reactionary forces like the Klan than to the emergence of working-class racial egalitarianism, carried into the volatile period ushered in by the war an attitude toward black workers that was, at best, ambiguous.

Workers of both races had chafed too long under the oppressive regime forged by district employers to let their wartime opportunities pass, however. Throughout 1916 and 1917, the "peculiar control" established by Birmingham operators since their victory in 1908 began to crumble under the weight of increased labor demand and massive out-migration. Taking President Wilson at his word, black and white Alabama miners hunkered down, fully intent on defending homefront democracy against the autocratic rule of the mine operators. Under such pressures the elaborate paternalist arrangement that only several years earlier had been lauded for delivering permanent industrial peace began to fray and ultimately disintegrate, as miners sensed that interracial unionism was once again a viable option. None were more surprised by the change of events than the operators themselves, whose faith in the stability they had established after 1908 seems to have been authentic.

Significantly, one of the most revealing manifestations of the miners' new confidence was their wholesale desertion of company-sponsored welfare work. A vexed TCI official complained that members of the company's brass band "were not willing to practice any more unless we would pay them for the time they had spent in rehearsal." Testifying before Governor Thomas Kilby's strike investigation in 1921, the Marvel mine owner Ben Roden—one of the district's most enthusiastic proponents of company welfare—recounted that he and his wife had together directed welfare work at Marvel since 1911. "When the union was organized there in 1916," he recalled disparagingly, "all of that work stopped." Attendance at company-run first-aid meetings dropped from 130 to only twenty-five, Roden fretted melodramatically: "Even the church work was confined to just a few men that would continue to come. [The union] seemed to take the place of religions and everything else; that was just their God."[65]

6. "People Here Has Come to a Pass": The 1920 Strike

AS THE PATERNALIST SYSTEM began to atrophy in the new context ushered in by the war, newly emboldened miners of both races engaged district operators in a series of hard-fought skirmishes. From 1917 onward the coalfields were convulsed in a state of continual industrial ferment. A chain of spontaneous, unofficial strike actions erupted at individual mines, leading eventually to coordinated action across the district and finally to Alabama's "most stupendous confrontation between labor and capital," the six-month-long coal strike of 1920–21. That bitter clash eclipsed in many ways the events of 1908 and, like the earlier strike, its outcome set the tone for labor relations in the Birmingham district for years to come. The conflict revealed unmistakably that despite its identification with progress and its promise of a break with the most repugnant elements of southern tradition, the paternalist arrangement and indeed the prosperity of the entire district rested firmly upon white supremacy and the continued exploitation of the laboring poor of both races. The entire edifice of "enlightened" management so carefully pieced together by the operators over the previous twelve years was abruptly abandoned during the strike in favor of a return to the methods that had secured their triumph over the UMW in 1908. In truth, little had changed in the intervening years; progress and reaction had coexisted at the heart of welfare capitalism all along.[1]

During the war of words that attended the descent into confrontation after 1917, mine owners and their allies in Birmingham's business and political establishment leveled two principal charges in their public campaign against the union. Genuinely convinced, perhaps, that the reform program instituted in the more advanced camps had "solved" the labor problem dogging operators in other fields, Birmingham mine owners could not comprehend the overnight resurrection of the union as anything other than a conspiracy brought in from the outside. Sloss-Sheffield executives, who had rather prominently resisted the trend toward company welfare, declined a government request to meet with union officials on the grounds that there existed "no disagreements [with employees] regarding wages, working conditions or living conditions." Hence the UMW's resurgence was attributed from the beginning to the work of "radicals," "carpetbaggers," and "outside agitators."[2]

Especially discomforting to architects of the paternalist order, however, was the apparent eagerness of black miners to forsake the ostensible privileges afforded them under that arrangement and lend their support to an interracial challenge. The Pratt executive and ACOA president Erskine Ramsay's frantic telegram to Labor Department officials in June 1917, in which he complained that UMW organizers were "stampeding . . . ignorant negroes" into joining the union, reflected the operators' perspective. A "citizen's committee" composed almost entirely of Birmingham's business elite would later complain, during the thick of the 1920 strike, that the tension pervading district labor relations was the fault of UMW officials. Accompanied by "a band of northern negroes and northern whites," union agitators had traversed the coalfields, operators complained, "for the first time bringing to the Alabama miners, of whom more than seventy percent are negroes, the news that they were underpaid and ill-treated." Discarding the benevolence with which it had adorned itself before the UMW's revival, little remained of the paternalist outlook after 1917 beyond the notion that rank-and-file miners—and black miners in particular—were incapable of independent thought or initiative. From there it was a short step to viewing all disturbances as the work of an alien fringe rather than manifestations of legitimate grievances.[3]

As evidenced by the UMW's revival, the conspiracy alleged in ACOA propaganda, which depicted otherwise content rank-and-file miners duped by unscrupulous "labor bosses," was not only false but directly contradicted by the actual course of events. While union officials proved themselves extremely reluctant to risk a decisive confrontation with district operators, ordinary miners continually forced industrial action, often acting independently of their elected leaders. Moreover, black miners, sensing along with their white counterparts a potential break in the operators' hegemony and buoyed by the heady optimism born of a newly reinvigorated racial militancy, proved unwilling to submit any longer to the old regime, placing themselves at the forefront in this new round of struggle. Far from being led along passively by white union officials, black miners displayed a fierce determination to secure economic and racial justice in the coalfields and, significantly, to force the UMW to live up to the spirit of racial egalitarianism formally inscribed in its constitution.

The enthusiastic reception accorded UMW officials during the early weeks of their reorganization campaign undermines the operators' claim that local miners were cajoled into organizing by outsiders. The opposite seems true: while district miners grew increasingly anxious for reorganization from 1916 onward, the national UMW balked, reluctant to commit itself to what was almost certain to be a protracted and costly war of attrition. "The miners at every mine in the State are ready to organize," a veteran Searles miner vowed in early June 1917, "and are only waiting the word and the proper authority to organize them." Looking back on the union's remarkable resurgence several years later, the ACLU field worker Mary McDowell confirmed that it was rank-and-file pressure rather than official initiative that drove the reorganization. "Miners as well as [local] officials say they sent one plea after another before the National sent any help," she recalled. The national UMW's prudent approach to the Alabama situation was also evident in remarks made by its vice president, John P. White, before a mass meeting of six thousand miners at Wylam in mid-June. Recalling his bitter experience during the 1908 strike, White warned his audience that strike action would only be undertaken "as a last resort" and emphasized his desire that miners and operators could reach a "peaceful adjustment" of their differences.[4]

While the UMW's preference for working out an agreement short of strike action and the federal government's interest in guaranteeing uninterrupted coal production might in other circumstances have combined to force a compromise, the operators' intransigence and growing restiveness among rank-and-file miners militated against such an outcome. Thousands of rank-and-filers obviously agreed with the socialist miner James Durrett, who urged *Labor Advocate* readers to "strike the iron while it is hot." "The time has arrived for your redemption," he advised them. "Brothers, if you let this opportunity pass it may never appear to you again." His sentiments were echoed by the black UMW member Jim Arnold, who complained that "there has been a hell of a lot of racket" made since the union was relaunched several months earlier, "but nothing has been done." When district miners met in convention on July 30, 1917, they issued an ultimatum to district operators: begin immediate negotiations on an agreement or face a district-wide strike beginning August 18. Among their demands, miners called for implementation of the eight-hour day, union recognition, a raise in wages, enforcement of the laws regarding checkweighmen, abolition of the subcontract system, and semimonthly paydays to be paid in legal currency rather than company scrip. The operators' flat refusal to even meet with District 20 officials only enraged miners further, prompting a series of spontaneous walkouts, including an unsanctioned walkout at Pratt Consolidated's Maxine mine.[5]

With the stage set for a potentially crippling confrontation, the labor secretary William B. Wilson, formerly a UMW official, intervened with a last-minute appeal to union officials to call off the planned strike action. "The uninterrupted production of the Alabama coal and iron district is of great importance [to] national defense," Wilson wired the District 20 president J. R. Kennamer on August 17. "Even a partial stoppage of work would be extremely injurious." Offering to mediate a conference between operators and the union on August 24, Wilson beseeched UMW officials to retract the strike call. "I cannot too strongly urge you to advise the miners to continue at work." On the same day, the federal government placed a full-page advertisement in the *Age-Herald,* with a personal message from President Wilson urging miners not to strike. Under mounting pressure and perhaps reasoning that a show of good faith would ultimately aid in gaining union recognition, local union leaders conceded Wilson's request, sending word throughout the district that miners should report to work on the 18th. The growing chasm between elected union officials and a membership increasingly willing to risk direct confrontation with the operators was evident in the lukewarm response to the back-to-work order. Thousands of district miners "simply refused" to return, while "many others reported to work reluctantly, feeling that their leaders had caved in to federal pressure and had nothing to show for it."[6]

Reports that mine owners across the district were continuing to gear up for a clash reinforced the suspicion among rank-and-filers that such concessions only encouraged the operators' obstinacy. The *Labor Advocate* reported a week earlier that TCI had "constructed a strong stockade" at its Edgewater camp and imported fifty gunmen, who were "parading around . . . armed to the teeth, surrounded by fully-filled cartridge belts . . . looking for trouble." In preparation for a strike, the company had reportedly brought "a number of negroes" from Tennessee "under the impression that they were to be employed in [steel plants]; but upon arrival . . . they were taken . . . and placed in the stockade, and told that they must go to work." District 20 officials complained that the company had "discharged more than five hundred men for joining the union" and was in the process of ejecting UMW sympathizers from company housing. Drawing upon their considerable expertise in manipulating racial divisions, DeBardeleben company officials orchestrated all-black "mass meetings." Under the watchful eye of company executives and race leaders like Oscar Adams, DeBardeleben's black miners passed resolutions vowing opposition to "any movement which seeks to frat-ernize white men and black men" and pledging to "stand by the company and work faithfully to protect it against any disturbment whatsoever." The importance of the gathering confrontation in the Birmingham district was not lost upon employers elsewhere in the South. Fusing traditional anti-unionism with the aggressive chauvinism increasingly evident across the United States, editors at the *Manufacturers' Record* reckoned that "if Alabama will handle this threatened strike, brought on by pro-German influences, as it should be handled, it will set an example to the nation." "No parleying, no dilatory tactics, no political-play of weak-minded men will avail in this hour," they cautioned.[7]

The calling off of the planned strike on August 18 only postponed Ala-bama's day of reckoning, however; it did little to soften the operators' nego-tiating position and even less to counter growing militancy among rank-and-file miners. In the days after the strike deadline had passed, the head of the federal Fuel Administration, Dr. Harry Garfield, managed to piece together a compromise that averted an immediate showdown but left miners extreme-ly dissatisfied. Under the terms of what became known as the Garfield agree-ment, union officials agreed to postpone implementation of the eight-hour day and forego any immediate wage increases. In return, operators pledged that they would implement the existing laws regarding checkweighmen and, while refusing to entertain recognition of the union, agreed in vague terms not to discriminate against union members in hiring or distribution of work.[8]

The reaction against the terms of this preliminary agreement from rank-and-file miners sent District 20 officials reeling. The *Age-Herald* reported that the pact "enraged" thousands of miners across the district. Workers at the Alabama Consolidated Coal and Iron Company reportedly walked off the job in protest on the day after its terms were announced, and deep disappoint-ment compelled outraged militants to organize a formal challenge to district leaders. A "dissident faction" of the union held a rally on August 28 to urge rejection of the pact, and the coalfields were rife with "growing talk of an in-surgent movement to unseat District 20's leadership." Local officials went into the ratification meeting planned for September 7 and 8 with little hope that the agreement would win the approval of rank-and-file delegates, and after a particularly "stormy session" in which they "strained every nerve to overcome the opposition of the men," Kennamer decided against offering the agreement up for a vote. He informed Dr. Garfield that unless the demand for an eight-hour day was incorporated into the deal he would be forced to authorize a district-wide strike.[9]

While awaiting a ruling on the outstanding issues from Garfield, Alabama miners conducted a "guerrilla struggle" during the fall of 1917, marked by spontaneous, unauthorized walkouts at various mines throughout the district. Though the agreement had formally prohibited blacklisting and reiterated the miners' right to elect a checkweighman from their own ranks, it provided for no mechanism to enforce these directives, and miners continued to complain that the operators were openly defying the government's instructions. Finally, on December 14, 1917, Garfield tendered his ruling, incorporating the demand for an eight-hour day into the contract and prohibiting reprisals against union members. From a situation where the UMW had been reduced to an ineffectual rump of long-time stalwarts before 1917, it had won broad support among miners of both races and managed—with the help of government officials—to force grudging, minimal recognition from district operators.[10]

Operators managed to extract a substantial concession in the settlement, however. The most significant aspect of the Garfield agreement was its attempt to curb the rank-and-file militancy that had become such a prominent feature of the UMW's revival. The contract stipulated that "nothing [could be] done to disturb the relation existing between employer and employee . . . and provided that employees joining any organization recognize the right of the company to insist that no employee shall use the company's time for any purpose other than that for which he is paid and that he must not interfere with the operation of the mine or knowingly reduce the output." To contain the kind of spontaneous walkouts that had been occurring with regularity around the district, the Garfield agreement provided for a permanent umpire to arbitrate grievances that could not be settled directly between the parties involved.[11]

Throughout the bargaining process, leading operators had maintained their refusal to recognize the UMW and would deny in future negotiations that their assent to the terms of the Garfield agreement implied union recognition in any form. ACOA members insisted that while they would meet separately with government representatives, they would not sit down to negotiations with any representative of the union. That they agreed at all to the Garfield pact is evidence of the defensive posture that they were compelled to adopt in the changed situation. A mere three or four years earlier they would have rejected outright any intrusion on their prerogatives. The exodus of skilled labor out of the district tempered the operators' intransigence considerably. "As long as the war lasted," Philip Taft has pointed out, "relations between the miners and mine operators were at least formally correct." Mine owners were willing, under the circumstances, to sit down with government negotiators. They grudgingly accepted the fact that wages would have to be raised to prevent a hemorrhaging of their labor supply. They were prepared to concede minor adjustments in conditions if it would guarantee them a steady supply of coal. But, with the exception of a handful of small operators, they drew the line at union recognition, the effects of which they feared would be felt long after the extraordinary situation brought on by the war had passed. Erskine Ramsay turned down a demand from the federal Commission on Coal Production to meet with union representatives, explaining that although the operators "[did] not wish to be regarded as declining to cooperate with you," they considered the request "tantamount to parleying with the UMW." Both sides would continue throughout the next four years to clash over ancillary issues, but the basic contest over whether the Alabama mines would be run on an open- or closed-shop basis would continue to drive tensions in the coalfields.[12]

The gulf separating operators and rank-and-file miners was such that the ink had hardly dried on the Garfield accord before violations were being reported in the coalfields. Kennamer informed district miners that the operators had accepted the terms of the agreement and that they should conduct themselves on that basis beginning January 31, 1918. Within weeks, miners at TCI had walked off the job over their employer's violation of the contract, only returning after being urged to do so by union officials, who assured them that the Fuel Administration would rectify their complaints. Despite these assurances, walkouts reportedly "plagued TCI and Republic coal mines over the winter of 1917–18." One operator, Ben Roden, later complained that miners at his Marvel mine seemed content to abide by the rulings of the government-appointed umpire only when he ruled in their favor. In both instances in which the umpire decided in favor of management at Marvel, "the men struck in violation [of the contract]," in one case for a week, until "union district officers finally went down there and made them go back to work." District officials now found themselves not only engaged in a war of nerves with the operators but being forced to assume the unenviable role of enforcing a contract that many of their members found intolerable.[13]

The fundamental dynamic driving miners and operators toward an open clash was barely affected by the Garfield accord, and both sides continued to gear up for a confrontation thought by many to be inevitable. Operators had agreed to the government's terms under duress and only on the condition that the agreement would not outlast the war. When the situation became more favorable, they clearly intended to disengage from any contractual relationship with the mine workforce and to attempt to reestablish the unilateral authority they had enjoyed before the war. One industry journal pointed out the importance of strong employers' organizations for the purpose of "combat[ing] trade unions" and warned that the "shaving down of wage scales [would] be necessary" in the postwar period. Miners were likewise dissatisfied with the arrangement imposed by federal officials, albeit for very different reasons. The new situation seemed to them an opportune moment to win better wages and conditions and to bar the possibility of a return to the status quo that had prevailed since 1908. They therefore resented delay on the question of wage increases, perceived by them as long overdue, and like their employers they awaited a break in the situation to assert their new-found strength.[14]

The first major confrontation after the war's end afforded both sides an opportunity to flex their strength. At its convention in September 1919, the national UMW voted to strike beginning November 1 unless the wage raises postponed during the war were forthcoming. Alabama miners joined the walkout, remaining out of work until the federal judge Albert B. Anderson issued a back-to-work order two weeks later. Protesting that their attendance at negotiations would amount to union recognition, Alabama operators boycotted the proceedings sponsored by the Coal Commission, later refusing outright to implement the 14 percent wage increase recommended by the agency. District 20 leaders had ordered a return to work on the understanding that the terms agreed to nationally would be applied in the Birmingham district and were outraged when Alabama operators not only ignored the recommendation but began a wholesale dismissal of union activists. Despite assurances from the attorney general A. Mitchell Palmer that he would "take prompt action against any operators who adopt such methods," the new round of blacklisting proceeded without any challenge from federal authorities, adding to tensions in the district.[15]

One ominous aspect of the operators' response to these new demands was a calculated turn to black labor. A Lehigh miner had reported as early as June 1916 that employers had added the phrase "MUST BE COLORED" to the employment listings appearing in local newspapers. Union officials protested that leading companies were utilizing medical examinations to disqualify white miners from employment, and their suspicions were confirmed by dissident company officials, who testified that the operators were taking on "only such whites as [could] not be gracefully thrust aside." A number of observers feared that the employers' careful manipulation of racial tensions, particularly during the steel strike in early 1918, had brought the district to the brink of race war. The Jefferson County sheriff J. C. Hartsfield penned a request to Alabama Governor Thomas Kilby for mounted machine guns to be used in the event of "labor strikes, race riots, etc.," warning that "in my twenty years experience . . . I have never known the situation to be just as it is to-day."[16]

Writing in the months leading up to the UMW's national strike of 1919, the NWLB field agent Edwin Newdick warned his superiors of a "bitter and dangerous alignment" that had been brought about as a result of the employers' "appeal to race prejudice." The Tennessee company, he pointed out, was considered "the paste-maker in this policy of discrimination . . . by which for years it has attempted to prevent the growth of the miners' union." "Employers now believe," Newdick wrote, that they had

> an opportunity entirely to disrupt this union. The plan is to get the negroes out of the united mine workers (negroes and whites are members of the same locals and officials of the mine workers state, work in complete harmony). A paper [the *Workmen's Chronicle*] was shown us . . . printed by a negro preacher in which was an article patently designed to arouse fear and prejudice in the negroes against the miners' union, referred to as "the white man's union." It was reported to the miners' headquarters that a wagon load of this issue of the paper . . . was seen dumped at one mine and that the paper is frequently distributed free throughout large sections of the mining district, if not at all the mines where the anti-union fight is being carried. . . .
>
> The readiness with which the miners' union could be disrupted once the negroes were out of it is apparent. Then, of course, negroes would rapidly and entirely displace white labor in the mines. . . . Labor leaders say that they feel that mine workers will not stand persecution much longer and that they would strike at once if their offices would sanction it.[17]

Newdick's apprehensions were seconded by H. P. Vaughn, a War Labor Board field worker who had lived in the Birmingham area for almost thirty years. Vaughn attributed the palpable rise in racial tensions in the district to the fact that "all of the Negro preachers had been subsidized by the companies and were without exception preaching against the negroes joining the unions." The "greatest bone of contention," he noted, was the activity of P. Colfax Rameau, the publisher of the *Workmen's Chronicle,* who toured the mining district under the operators' sponsorship, "holding meetings for the ostensible purpose of talking health and hygiene to the workers" in "halls and other convenient auditoriums on the company's property" to which "union men are not allowed to go." Vaughn estimated the *Chronicle's* circulation to be "very large" and confirmed that it was distributed "free of cost to all the colored men in the mines and mills."[18]

Racial antagonism was thus injected as one of the central features in the new round of confrontation from early on, presenting UMW officials with prospects of a repeat of the calamitous events of 1908. Caught between the employers on one side, whose public pronouncements continually emphasized the union's threat to white supremacy, and the operators' black middle-class allies on the other side, who warned black miners to steer clear of the "white man's union," UMW officials were hard-pressed to construct a policy that could hold out to black miners the potential for tangible gains in their working lives without invoking the wrath of the guardians of white supremacy. Like his counterpart, Vaughn recalled that the UMW "has been taking in colored men ever since it was organized, and their officials are practically divided equally between white and colored" but noted union officials' concern that the operators were "trying to bring on a race antipathy in a situation where an industrial fight is the one most of them are concerned about."[19]

The reflexive response of District 20 officials to the charge of racial subversion was to deny outright any intention on the part of the UMW to upset the southern social order. In part this expressed the genuinely conservative racial outlook of leading white officials, who seem to have been fully committed to winning equal pay and conditions for black miners but who had no intention of leading a broader fight against white supremacy. Many white miners and officials grasped profoundly how district operators had effectively resorted to "race-baiting" in the past and were astute enough to perceive that their wartime hiring policies were motivated by antiunionism rather than newfound racial preferences. At the same time, they feared being overwhelmed by racial demagoguery and, at a time when the charged atmosphere of the war years had revived racial paranoia throughout the South and breathed new life into organizations like the Klan, calculated that success on the industrial front hinged on their ability to deflect the taint of racial subversion. UMW officials thus determined that the most effective response to the employers' race-baiting was to downplay the race question altogether by insisting that theirs was an industrial struggle that did not impinge on race relations. One of the central questions posed by the experience of the UMW over the next four years is whether such a distinction was possible at a time of acute racial tension and in an industry whose workforce was more than 50 percent African-American. Certainly the operators were unimpressed by the UMW's logic and perceived the challenge to their industrial supremacy as an assault on the broad foundations of southern society.

The operators' open defiance of the U.S. Coal Commission, their increasingly flagrant resort to racial manipulation and red-baiting, and the palpable militarization of the coal district all pointed to the increasing likelihood of large-scale confrontation. The *Labor Advocate* reported in early 1920 that a campaign was underway to depict "the skilled trades workers of the Birmingham district [as] red-eyed agitators and anarchists," to encourage the belief that "they embrace the soul-destroying doctrines of Russian bolshevism." "One class of producers who are inclined [*sic*] in this insidious attack are the coal miners," the editors noted, "especially those of Walker County." For their part, union miners seemed no less determined than their employers to press forward. Meeting in convention in April 1920, district miners adopted the so-called blue book agreement in an attempt to consolidate the tentative recognition they had secured through the Garfield accords. Union officials instructed members that wherever conditions were favorable, "the employer should be requested to sign the blue book agreement, and in the event of a refusal a strike was to be called."[20]

By this time thirty-five smaller operators had voluntarily signed union contracts, but the leading companies, supported by a majority of the smaller, independent operators, refused to enter into negotiations and attempted instead to actively prevent the growth of the union. While rank-and-file miners proved more eager than their officials for a showdown, the passivity that had set in after numerous false starts over the preceding period hampered the remobilization of the UMW's membership in 1920. J. R. Kennamer, sharply criticized earlier for his moderation, found during his forays into the coalfields in the spring of 1920 that many miners who had earlier been anxious to strike were now holding back from active involvement in the union. In a speech at Oakman, Kennamer (a Baptist preacher) rebuked those who had "backslid" since 1917. "Some of the miners in this district remind me of the Baptists," he told his listeners. Where previously, having "taste[d]" the "excitement about the UMW," miners had "flocked" into the union "by the hundreds and thousands," their disappointment at not being allowed to strike had caused many to "[fall] from grace."[21]

Strikes nevertheless began to burst forth across the district in the spring of 1920, and by midsummer five thousand miners were out on strike. When, in July 1920, Sloss company officials prevented a UMW organizer from speaking to an assembly at Coal Valley, the *Labor Advocate* reported that the refusal of "mines along the Southern Railway [owned by 'the Sloss Company, Pratt Consolidated, and Brookside Company'] to recognize the union" had caused "considerable dissatisfaction . . . among those workers," and UMW officials warned that "unless these companies recede from the policy of non-recognition they [the union leaders] will not be able to keep the men at work." A Labor Department official reported in mid-July that the situation was "growing more acute and the miners and operators further apart than ever." Evidence of this could be found in the half-page advertisement taken out by the Galloway Coal Company president F. N. Fisher, who vowed to continue operating on an open-shop basis and expressed his determination "never to recognize any organization which seeks to deny any man, white or black, the exercise of [his right to work through a strike]." The DeBardeleben supervisor Milton Fies wondered privately whether operators should attempt to undercut union support by "put[ting] in a [wage] raise before the union takes credit of forcing [us] to do it." Union leaders warned federal officials that they fully expected "a fight to the finish over the issues involved." "Gun men are being imported into this section," a federal agent declared ominously, and "the men are being intimidated in every way. The situation is bordering on riot."[22]

With little room for compromise, District 20 officials faced the choice of organizing an effective district-wide strike or giving way before the steady onslaught against modest gains won since 1917. The "policy of sporadic strikes of selected mines was failing," notes Richard Straw, and the *UMW Journal* reasoned that officials had to either "call a general strike and fight for the rights of the miners" or "surrender . . . and leave the state without the vestige of an organization." John L. Lewis, the president of the national UMW, authorized a strike to commence on September 7, 1920, in order to "compel the operators to meet . . . and settle the questions at issue between the mine workers and the operators." Setting the tone for the bitter dispute that would follow, the ACOA committeeman James Bonnyman responded by pointing out that "the operators of Alabama have and always will refuse absolutely to treat directly or indirectly with the United Mine Workers of America."[23]

About half of the district's twenty-six thousand coal miners responded immediately to the strike call, with five thousand still working under union contract and another eight thousand defying the order and remaining at work. By the month's end, Van Bittner, dispatched to the Alabama fields to coordinate the strike on behalf of the national UMW, reported that the district was in "splendid shape." After convening a special district convention in Birmingham on September 22, Bittner wired John L. Lewis to inform him that the "greatest strike in the history of Alabama is now on in full force." Two days later he issued a statement vowing that every nonunion mine in the state would be shut down within a week. "This will be no front porch campaign," he vowed. "It will be conducted in the coal fields." In a letter to the UMW vice president William Green, Bittner reported that "hundreds of men are leaving the mines and joining our ranks every day." "The sentiment of the business men of Birmingham is bitter," he admitted, "but in the mining and farming communities the sentiment is generally favorable." Failure to win the strike, Bittner acknowledged, would deal the Alabama UMW "a blow from which [it] will not recover for many years to come."[24]

Impressive early strike figures, which seemed to justify guarded optimism in the union's ranks, understated one crucial weakness in the UMW campaign. Among the district's major operators—TCI, the Alabama Fuel and Iron Company, Pratt Consolidated, and Republic Iron and Steel—production remained at or near normal levels. The battleground for the duration of the strike would shift to these heavily capitalized and closely guarded operations. These leading employers in turn provided the key personnel for the ACOA's "Co-operative Combat Committee," which directed the operators' campaign and seemed determined to block any settlement short of total victory over the union. They were supported from the outset by two key constituencies: Birmingham's influential business community, led by individuals who had taken part in organizing the Vigilantes several years earlier, and the city's black middle class, which in the coming months would play a critical role in bolstering the operators' position.

The aggressive intervention of these two groups in the early days of the strike aided the operators in isolating the UMW and stirring up public antipathy against the miners. Oscar Adams hosted a large, mostly black meeting at Birmingham's Metropolitan AME Zion Church during the second week of the strike, at which he denounced the UMW as "a pack of 'I.W.W. Bolsheviks' and 'agitators' bent on violence." Under his editorial direction, the *Birmingham Reporter* kept up a steady denunciation of the UMW, accompanied by glowing accounts of conditions in the coal camps and reports that black strikebreakers were "breaking all records in coal production."

Adams, Rameau, the Republican Club chairman W. B. Driver, and other prominent race leaders had been centrally involved in efforts to dissuade black miners from joining the UMW since the beginning of the war. Following the lead of the operators, they had stepped up their activity in the early spring. "The leaders of the race almost to a man," reported the black press in 1917, "have with one accord advised their people to stay out of the white man's union." At the Bonnyman Coal Company's Coalmont mine, where blacks comprised more than 80 percent of the workforce, Adams had been instrumental in organizing a Civic League as an alternative to the UMW. The explicit antiunionism of their message was evident in the remarks of Adams's close companion in this work, the Reverend A. C. Williams of the influential Sixteenth Street Baptist Church. Williams advised black miners that if they "must organize, let [them] organize [independently of whites], and then not to antagonize capital, but to . . . promote efficiency, and to secure [gains] through the co-operation and sympathy of the employer."[25]

The formal launching of a district-wide strike inevitably heightened tensions between the black middle class and black miners and between race leaders and UMW officials. Adams seems to have been taken aback at the widespread support that the strike elicited among black rank-and-file miners, publishing an editorial on September 11 in which he regretted that there were "so many innocent, misled and misguided Negro men involved in this unpleasant mining situation." The UMW's embrace of interracialism pushed race leaders to adopt a somewhat bizarre defense of the color line. Adams denounced the union for "insinuating that Negro miners and white miners are social and industrial brothers. They are neither." Eventually race leaders would join the operators in condemning union interracialism as a policy that defied "the plans of human society." Rameau would go so far as to urge a special session of the state legislature to pass a law "prohibiting Negroes from joining any white man's organization, fraternal or industrial," a measure that he claimed would "do more to help the Negro . . . than anything that has been done . . . since the signing of the [Emancipation] Proclamation."[26]

Vilified by district operators for their "history . . . of associating the black man on terms of perfect equality with the white man" and by black elites for holding out a false promise of industrial brotherhood, UMW officials responded early in the strike with a tactless attempt to drive a wedge between race leaders and their own African-American members. Speaking before a crowd of black miners at Sayreton in mid-September, J. R. Kennamer condemned the efforts of "influential [black] leaders" who were "opposing the union and trying to preach to colored miners to stay at work." In an uncharacteristic departure from his usual moderation, and one that evoked the most offensive elements of southern racism, Kennamer suggested that if blacks at Sayreton "have such leaders among you . . . , you should take them out and hang them by the neck."[27]

Kennamer's comments, though apparently applauded by black miners at Sayreton, elicited widespread condemnation in the mainstream and black press and forced the UMW onto the defensive. Relations between black miners and race leaders with close ties to the coal companies had been deteriorating for some time, however. With the eruption of industrial unrest in 1920, these tensions exploded into the open. Earlier that year a black minister who was "said to have tried to persuade miners not to strike" was found dead and presumed to have been murdered by union men at Praco mines. Prominent whites detected in Kennamer's speech a bold attempt to "incite the colored miners to violence and turn them against their conservative leaders and preachers." Adams took the opportunity to issue an editorial linking "Lynching, Striking, and Rioting," though ironically the strike's only apparent lynching would be carried out by members of the Alabama National Guard.[28]

Although the black establishment's attitude toward the strike would continue to pose a major problem for UMW organizers, union officials would in the future restrain themselves from such demagoguery. Speaking before a racially mixed group of miners a month later, Van Bittner singled out a reporter from one of the black newspapers, reminding his audience that "the [UMW] has put dollars in the colored men's pockets here in Alabama where he has taken dollars out." Emphasizing that he "did not care an eternal damn what [a miner's] color is, so long as he fights," Bittner challenged the journalist (possibly Adams) to "a joint debate before the colored miners of Alabama" to "let the colored men decide for themselves whether they want his policy or the policy of the UMW." The importance that the UMW attached to reaching black miners is attested to by their distribution of an "Appeal to the Colored Mine Workers of Alabama," a pamphlet that contrasted the UMW's pursuit of industrial equality with the operators' hostility to "associating the black man on terms of perfect equality with the white man."[29]

The first major incident in the strike came less than two weeks after its commencement, on September 16. Relations between mine management and the workforce at Corona Coal Company's Walker County mines had been strained since the early summer. As early as June 1920, the Townley mayor E. C. Ellison had written Governor Kilby to complain that the mine superintendent Leon Adler and an armed, "organized band" of Corona mine guards "had taken the situation in their own hands" there, assaulting union sympathizers and town officials in broad daylight. Ellison reported that Adler's men had "waylayed" the town marshal on June 21 and "with their hands on their pistols subjected him to every vile epithet that could be applied," while loudly proclaiming that "by god he [Adler] was running this place." Having received requests for protection from both whites and blacks at Townley, the mayor requested that the governor dispatch a "small company of the state militia" before Adler and his "organized gunmen . . . start blood to flowing." A month later, Ellison forwarded a second urgent appeal after Adler and his "gang" intervened to prevent a legal miners' mass meeting in the town square. Dozens of miners appeared on the scene armed with shotguns, and a pitched battle appeared inevitable. Confrontation was narrowly averted, but the local sheriff's complicity in the events left a sense among citizens that law enforcement officials were "unduly influenced by the Coal Companies."[30]

The confrontation at Corona finally came on September 16. Tensions there had escalated since the beginning of the strike, and after noting that "striking miners with pump guns, shotguns, pistols, and dynamite" had been observed "patroll[ing] the railroad tracks and public roads" the day before, the *Birmingham News* had ominously predicted, in its morning edition, that "bloodshed is likely by the end of the day." Birmingham's major dailies would later report that Adler and thirty of his men were ambushed by fifty armed miners while attempting to prevent them from "marching on Patton Mines for the purpose of forcibly stopping work." The labor press disputed this account, however, insisting that Adler had been shot and killed (along with Earl Edgill, one of his men) within a short distance of the UMW union hall at Corona. Union sympathizers reported that Adler had ordered his men to shoot into the hall, waiving aside objections from his own ranks that they should "not go messing with [union miners] in their hall." Adler allegedly insisted that he could "take a pocketful of rocks and run every white man away, and then take a stick and run all the negroes back to the mines." Besieged by the operators in the wake of Adler's death, Governor Kilby declared martial law, dispatching troops to the mining district the following day. Senator George Huddleston, speaking before a mass rally in Birmingham several days after the shootout, warned that the situation in the coalfields was growing increasingly volatile. "The district trembles," he cautioned, "trembles on the verge of civil war."[31]

The bellicose mobilization of Birmingham's business community against the strikers deepened the polarization that rent the district in the wake of Adler's death. Open calls for a revival of the Vigilante organization, which had played such a prominent role during the 1918 steel strike, were issued at a boisterous meeting of the city's Kiwanis Club (of which Adler had been a member) within days of the incident. At that meeting, Russell Hunt, a high-ranking Sloss-Sheffield official, referred to the newly arrived UMW organizer Van Bittner as "Von Hindenburg," and a prominent banker, T. O. Smith, called for a return to the private regulation of justice. "We stand for law and order," Smith reminded the gathering, "and if we have city and county officials who cannot enforce the law, the Vigilantes will." Assuring his audience that "we do not believe in using force except as a last resort . . . after the constitutional authorities fail," he vowed nevertheless to "be one of one hundred men to give up my job . . . and shoulder a gun if necessary." Their reaction to the Adler incident would prove to be the first in a long series of business community interventions in the strike. Though the evidence is scant, it is clear not only that private citizens played a role in policing the strike but that Governor Kilby authorized the deployment of "one hundred or more good citizens" as a "volunteer" detective force.[32]

The perception by miners at Townley and elsewhere that public law enforcement officials were colluding with mine owners and their deputies seems warranted, and widespread collaboration between private gunmen and public officials seems also to have set an important precedent for the role that state troops would play in the dispute. In the months leading up to the strike operators had devoted considerable attention to purging the coalfields of officials deemed overly sympathetic to the union. In July 1920, for example, the coal company attorney Borden Burr wrote Governor Kilby to inform him that his client (the operator Ben Roden at Marvel) was planning to import strikebreakers and to demand that the local sheriff, considered a UMW sympathizer, be removed.[33] While no record of Kilby's response survives, we do know that less than two weeks later an unnamed deputy sheriff at Marvel was "asked to come to the office of the mining company on some pretext, [where] he was seized by the guards, disarmed, and threatened with physical violence if he did not immediately leave the camp." The *Labor Advocate* explained that this deputy had "refused to be a party to the practices of the mine guards employed by the coal company and was in the habit of giving the miners a square deal." In a similar attempt to guarantee a pliant police force, the operator W. N. Denton wrote Kilby complaining that the Jefferson County sheriff J. C. Hartsfield's "continued protest against the presence of troops" suggested that "the gentleman in question would like to leave the way open for intimidation and a 'reign of terror.'" As these incidents suggest, coal operators did not automatically accept public jurisdiction over the mining camps or adjacent towns. Amicable relations prevailed only where legal authorities demonstrated a willingness to cooperate in putting down the strike. Leniency toward the strikers, however, or even genuine neutrality was perceived as evidence that an individual lawman was unfit for service in the coalfields.[34]

The logistics of military deployment did little to assuage suspicions that the troops were there to break the strike. Upon their arrival in the field, state troop commanders typically placed themselves at the direct service of local company officials. In late September, when a commander of the state militia at Blocton approached miners there to inform them that they "could not attend any public gatherings," he was accompanied by the TCI superintendent James Stewart. When asked about religious services, the commander explained to miners that they were prohibited from attending, as "not more than two people would be allowed to gather at any place." Similarly, the leader of a squadron at Benoit informed miners there that "the churches would be closed up, the doors being nailed," and when they were reopened more than a week later, a notice co-signed by the commander and G.S. Bissell of the Benoit Coal Company proscribed the use of church facilities for strike-related purposes.[35]

Even where company officials were not directly involved alongside state troops the effect of martial law was to deprive union miners of any means of obstructing coal production or spreading the strike. Though alarmed by the aggressive campaign being waged against them in the local press, union officials were far more concerned in the first few weeks of the strike that military intervention would frustrate any possibility of making the strike

effective. "The soldiers have dispersed all mass meetings that we have attempted," Van Bittner informed national UMW headquarters, and "refuse
to allow us to hold any meetings except the regular business meetings of the
local unions." Acknowledging that "the strength of our organization has always been public mass meetings," he predicted that "if we would be allowed
to hold [them], there is no question but that within two weeks we would close
down every mine in the state."[36]

This the military seemed determined to prevent. The orders stipulated that
the military command had to be informed in advance of any meeting of three
or more miners, and throughout the coalfields these orders were interpret-
ed to prevent not only formal union meetings but religious services and
meetings of the camp fraternal orders as well. When the black District 20 vice
president Joe Sorsby attempted to address an assembly of miners at Hillsboro
in Shelby County, he was stopped "in mid-speech" by a "lieutenant who said
that he had to confine his remarks to local organizing work." Occasionally
such heavy-handed intervention backfired on the military. When a corporal
at Searles, in Tuscaloosa County, tried to prevent Sorsby from speaking on
the grounds that "no outsider could speak at said meeting," four hundred
miners "decided unanimously to quit work" (there had previously been "no
strike or cessation of work at these mines"). Outraged, they vowed to remain
out until assurance that "their rights [would not] be trampled in this way"
had been secured. A similar stoppage occurred at Brookwood after the military "interfered and broke up a meeting" there.[37]

Throughout the coalfields miners bristled under military control. An
Anniston miner wrote Bittner in late September to support the UMW's effort to "create the Lord's Prayer on Earth (they [*sic*] will be done on earth as
it is in heaven)" but protested that the military had presided over his local
union meeting the previous week, at which "we was not permitted to speak
of anything . . . outside our own [local affairs]." From Carbon Hill came the
complaint that "we are under Martial Law here and the auditing committee
asked permission to meet and audit the books which the [captain] refused
to grant." Miners at Sayreton found their local meeting limited to ten minutes. UMW members at Brilliant wrote Governor Kilby to ask, "Is it a fact
that we can not have a Masonic communication here . . . without the presence of a soldier? What law are we under?"[38]

From the first weeks of the strike, miners and their officials perceived the
power of the colossal forces ranged against them. Although in his public statements Van Bittner exuded an air of confidence about the UMW's ability to
persevere and secure a closed shop, privately he acknowledged that "the situation here is the worst I have ever been confronted with. Every force of

government and everybody else not directly connected with the trade union movement is against us." Bittner's optimism about the strike's prospects and his sober assessment that the union could not prevail on its own are evident in a letter he forwarded to national headquarters at the end of September. "I feel confident," he told UMW executives, "that if we can continue the battle that the least we can get is some sort of government intervention and a settlement . . . which will protect our organization in this field."[39]

Without ignoring the war of words developing in the local press between themselves and their adversaries, UMW officials recognized early on that the strike would be won or lost on the picket lines. They would either succeed in shutting down the mines or the operators would maintain levels of production sufficient to brush aside the miners' demands. Here the operators enjoyed a number of distinct advantages. The military presence was the most obvious one: "promptly upon the coming into the district of the troops," a report sympathetic to the operators acknowledged, "coal production increased." Moreover, since 1908 Birmingham district operators had built into contracts with their clients a strike clause that provided for a renegotiation of coal prices in the event of industrial unrest. Observers remarked that the strike afforded a number of prominent operators an opportunity to terminate arrangements that had locked them into wartime prices: they therefore welcomed a limited strike. More importantly, though, the long delay between the UMW's reorganization in 1917 and the strike call in 1920 had allowed leading operators ample time to develop emergency plans.[40]

In particular, great care had been taken to begin the transportation of strikebreakers into the district from before the commencement of an all-out strike. The *Birmingham News* reported during the first week of the strike that the operators had expressed confidence that "in a short while several thousand green hands will have been brought into the district and many of the men striking will return to work." Although it is true, as Richard Straw has written, that there was at this time "no indication that these 'green hands' were blacks," reports from throughout the district suggest that African Americans made up the overwhelming majority of those imported to break the strike. "They are getting negroes in here every day from Birmingham and South Alabama," a UMW official at Porter informed district headquarters. "There are many men being transported here from the southern part of the state," the veteran black miner D. J. DeJarnette wrote to Van Bittner. "It is the Negro they are hitting on, my race." And although in one sense it seems unremarkable that the operators turned to the Black Belt for strikebreakers, they were no doubt encouraged by the possibility that such a solution would stir up racial animosity within the ranks of the UMW.[41]

Perhaps the most eloquent testimony to the vigor of coalfield interracialism is the resilience of class solidarity in the face of such flagrant provocation. An incident that occurred in Walker County over the previous summer illustrates how the polarization between miners and their adversaries could blur the district's otherwise inviolable racial lines. Following an altercation at the Carbon Hill commissary between Sheriff Ben Barrett and a black union miner who was apparently "mak[ing] some threats among other Negroes as to what would happen if certain men went to work," the white UMW member William Hicks "started after" Barrett and his deputy, Oliver McDade, shooting both of them fatally. Hicks was then pursued by a posse and "finally riddled with bullets." Similar violations of racial protocol occurred on a regular basis throughout the strike: black strikers attacked white and black strikebreakers; white strikers broke with politicians of their own race and physically defended black union organizers against company and state repression; black UMW members spurned the appeals of "race leaders" in favor of those from UMW organizers, both black and white. If such actions seem unremarkable by modern standards, they bore a far more troubling salience for the operators and their allies and seemed to portend the collapse of the foundations of southern society.[42]

Two aspects of the UMW's efforts to prevent strikebreaking seem especially noteworthy. First, it seems clear that wherever union members enjoyed access to scabs they could be easily dissuaded from breaking the strike. Second, the hostility that strikers displayed toward scabs seems to have contained almost no specifically racial element. In many cases white miners seemed sympathetic to the plight of black strikebreakers, acknowledging that it was their bleak predicament rather than any innate proclivity to scabbing that had cast poor blacks as the easy prey of the operators. By contrast, miners of both races exhibited far less tolerance toward strikebreaking by local farmers, who were far more likely to be white than black.

The union's ability to turn away strikebreakers rested upon two pragmatic considerations: physical proximity to strikebreakers and access to minimal financial resources to provide necessities such as food or transportation to any they succeeded in dissuading. Many of those transported into the district were unaware that they were being imported to break a strike and were bewildered (not to mention frightened) at being met at railroad depots by throngs of heavily armed union miners. Union locals throughout the district reported to Van Bittner that wherever they could secure a hearing among these men they were remarkably successful in convincing them to honor the strike. "They gets a lot of colored fellows in here but they don't stay long," a member of the Parrish local informed Bittner. Roebuck miners reported similarly that they "turne [sic] back crowds of them when we tell them how it is hear [sic]." From Jasper came word that "labor agents are bringing negroes into this vicinity in great numbers claiming to be building a railroad; all we are able to get in touch with are more than anxious to leave when they find out they are going to work in a mine."[43]

Gaining a hearing among strikebreakers frequently called for creative tactics on the part of UMW locals. "We cannot get around to them," members of the Blocton local complained. "They work the Poor Devils seven days a week and they stay in at nights." One miner, R. F. Hancock, penned a letter to Governor Kilby after the Birmingham press published P. Colfax Rameau's assertion that black miners were "satisfied and wanted to work." "As to the amount of men that wants to join the [UMW]," Hancock wrote, "I can say that there is camps . . . guarded for the sole purpose of keeping the negroes from joining the union." He recalled a recent visit to TCI's Edgewater camp and asked "if these men [don't] want any union why does this great company guard them inside of this wire fence that they have around that camp?" At Searles the black vice president of the UMW local defied management orders to evacuate company housing and was warned by the superintendent that he "was the sole cause of the colored miners not working, so I am up to my last notch." The UMW local at Glen Mary mines decided that in order to bring out strikebreakers there, they would send one of their own members into the mines. William Bradley, a native Tennessean, volunteered to go to work among a dozen scabs, and over the course of a single working day he "brought all the scabs or strikebreakers up out of the mine and sent them out of town." Bradley felt confident that with the union's consent he could "do the largest majority of them that way." J. Frost reported from Parrish that the union was growing despite the fact that "they won't let us have no meeting." "We got some of [their] best driver[s] away from them yesterday. The companies make like they loads so much coal a day, but that's just all a mistake . . . they moved some more scabs in [to company housing], and they worked a day or two, and then joined the union."[44]

Thrust into the middle of a fierce labor dispute but obviously uncommitted to weathering the storm alongside the coal operators, strikebreakers seemed more than willing to desert the mines. In mid-October Bittner expressed his certainty that "these men are scabbing because they don't know any better; they will be a part of us before many more days are over." "We want these men in the union," he reminded strikers. "None of us were born with a union card in our pocket. . . . Get these men into your union and treat them the same as you would any other union man." One important factor facilitating the UMW's efforts to minimize strikebreaking was the scabs' harsh treatment at the hands of the operators. When an explosion killed a number of strikebreakers at Parrish, District 20 sent a check in the amount of five hundred dollars "to relieve the immediate suffering caused by this explosion" despite the fact that "these men were not members of our organization." After a young strikebreaker at Masena was crushed under a mine car, strikers expressed their sympathy for the boy and his family. "We tried our best to get these boys [the injured miner had two other brothers scabbing at Masena] not to SCAB but failed," a union miner assured Van Bittner. "So you see the results. Poor little boy won't never be much good if he gets well." Relations between the operators and their new labor force at Brookwood mines deteriorated so drastically that by late November scabs themselves launched a strike inside the stockade erected to seal them off from contact with the UMW. At Parrish bosses fired a black strikebreaker after he refused to take part in an attack on the UMW hall there and joined the union instead. One of his coworkers also joined after refusing to work a ten-hour shift, and several days later the convicts at Parrish launched a strike of their own.[45]

The most difficult obstacle to ridding the camps of strikebreakers proved to be the increasing difficulty of providing them with food or transportation out of the coalfields. The strike committee at Jasper noted that "all [the strikebreakers] ask is railroad fare away from here to Birmingham" and prevailed upon Bittner to advance funds for such a purpose, believing that they could thereby "get rid of at least seventy-five percent of the men they bring in." From the first weeks of the strike, however, union funds were stretched thin, a deficiency that would make itself felt in every aspect of the strike effort.[46]

The situation at Marvel became so acute that the local's financial secretary wrote Bittner to complain that "the president of Local 3275 will not consent for us to organize the scabs, [because] he don't know whether the district will feed them or not. If . . . District [20] will feed them we will try to organize Marvel." The presence of several dozen Italian strikebreakers at Marvel posed a unique problem for the local there. UMW officials wrote Bittner after being approached by Italian union supporters with the unusual request that they be paid in cash rather than standard union rations so that "they might buy such things as such people live on." Strikers apparently believed that they could thereby "get all the Italians to quit work," among whom were about eight "experienced coal diggers." The union responded that they could not fulfill the request: "Furnishing macaroni and cheese for the Italian members . . . is the very best that we can do." At Brilliant strikers found themselves locked in a seesaw struggle with operators, who were continually intimidating black miners to force them back to work. As long as black miners lived under company jurisdiction, local union officials believed, the company would have the upper hand. "The Superintendent has taken a trip to Birmingham and told some of the striking negroes that they better be at work or be gone when he returns or he wood [sic] take a stick and beat them out of the houses." Local leaders prevailed upon the union to send tents: "If we can't get somewhere to move these nigroes [sic] we can't hold them." Union officials at Porter faced the same dilemma: "I am sure I can get some of these negroes out [but] if I have no place to put them there damd [sic] thugs bluffs them back to work."[47]

To be sure, a strong element of intimidation pervaded confrontations between the UMW and strikebreakers. An article appearing in the *Advance,* one of several small newspapers sympathetic to the union, in early December outlined the method by which scabs might be dissuaded from working. Success in other mining fields, the paper reported, "had usually been accomplished by union men forming bodies at a given point and marching in certain cases on a camp where scabs are working. They advance on the place in group formation in order to overthrow the company gunmen and guards . . . [and] advise the non-union miners of the injury they are doing. . . . They then invite the non-union men to cease work and leave town, a request that is usually said to be granted." While it is difficult to ascertain how many such "invitations" were parleyed in the course of the Alabama strike, this sort of direct action is precisely what the operators and the military intended to suppress. The operator Ben Roden reported that strikers at Marvel "did everything they could" to stop the scabs. "They would meet the trains" with "a mob of one hundred and fifty or two hundred men . . . every morning and night, and any men who would come in for work they would follow them around, just like they were some kind of side show exhibit, and jeer at them, and talk to them; and we just had to guard those men day and night."[48]

Naturally, there were limits to the UMW's patience, and as the strike wore on the union's early perspective of winning converts gave way to frustration and the conviction that those remaining at work had consciously chosen to side with the operators. In place of the generosity that had characterized Bittner's early speeches came a declaration that "the scabs had been given sufficient warning." "If a few of them are caught in the rush and suffer the effect of staying in the camp too long it will be their own fault," he warned. At a special convention on December 10, district officials set up local "picketing committees" charged with doing "everything in their power to prevent men from going to work in the mines, and . . . to have those who are now working join the strike."[49]

Deep antipathy marked relations between UMW members and committed strikebreakers. At Marvel, Roden reported, strikers and scabs "did not speak. Some of them were living right side by side and their children would throw rocks at the other children and there was just practically an armed warfare there." Management loaned out thirty shotguns to strikebreakers who "were . . . afraid to go to work and leave their wives and children . . . unprotected." Masena strikers took some comfort in the predicament that scabs there found themselves in when the mine went bankrupt. The strikebreakers "are greatly grieved here today," a union member reported, "as the check issue[r] is telling them . . . his hand is paralyzed [*sic*] now and cain't write them any." Conditions had become so dire that white scabs took up residence in the "colards'" boarding house, where "they claimed they had more to eat . . . than they ever got at the whites," a situation that union miners, themselves under attack for breaching the color line, found amusing: "That is there [*sic*] social equality they have been throwing at us . . . ever since the strike in 1908," one striker quipped. The company's rough treatment of scabs, he believed, would provide "one of the greatest lessons to those [unionists] that are weak." "Ha ha," the striker scoffed, "some of the old heads that has fought us so hard down here . . . are pouring out an honest confession to redeem their dirty conscience, but I for one can never forgive."[50]

Fervent hostility toward scabs became increasingly evident as the strike wore on; the press reported numerous incidents where individual strikebreakers had their homes dynamited or were shot by strikers, and wherever they could find a way around the military, as in 1908, strikers ambushed the trains carrying scabs to and from the coalfields. Nevertheless, a Birmingham grand jury would later find, after a "review of the conduct of the miners in this strike," that with the exception of "some few [who] are to be expected to step beyond the bounds of law and order," "a greater number . . . stood for the observance of law and order than at any time in the history of similar strikes in the past."[51]

It is significant that in spite of the operators' heavy reliance upon blacks as strikebreakers, virtually none of the sources that reflect union miners' hostility toward scabs show any evidence of racial antipathy. Both black and white unionists displayed less patience for the (mostly) white northern Alabama farmers who crossed their picket lines. Union members who felt some empathy for the plight of those being shipped in "just like cattle in box cars" apparently believed that the farmers had no excuse for their actions and that many were taking advantage of the situation to live off of the UMW. "Most of [the farmers] went to . . . work and what few did stop to talk to us said they would have to go to work if the [UMW] didn't feed them," Blocton miners wrote five months into the strike. When mines began laying off in mid-January, "lots of these that went to scabbing . . . are trying to get on the [union ration] list." "The farmers has taken advantage of us in every way in this strike like they did in 1908," they wrote. "Lots and lots of them made big crop . . . and the union fed them." Resentment against the "farming element" became especially pronounced as hardship set in. A Quinton miner who had joined the union in 1899 wrote the UMW organizer William Turnblazer to warn him that the crowds cheering his speeches in the mining district included many farmers who "never did, and never will go down into [the] Mines." "Some of them were there because they were with us in our Fight, but most of them were there as frauds for the sole purpose of [getting] on the Bread Wagon FREE, and do not give a D— for nothing else." This veteran advised Turnblazer that the union locals needed a "real shake up, [during which] the farmers and drones [should be] kicked out."[52]

The notable lack of a racial element to the strikers' anger is worth examining. Partly it reflects a realization among white UMW members that a focus upon the racial composition of the strikebreakers would play into the hands of the operators by splitting the union along racial lines. No doubt it also derived from the recognition among whites that black strikers were sacrificing much to defend the union. The determined role that black unionists assumed in prosecuting the strike (inevitably juxtaposed to white farmers' role in undermining the UMW effort) made it difficult to sustain racial generalizations about blacks' proclivity for strikebreaking. One of the most dramatic illustrations of black miners' rejection of paternalism came at Dora, where a small group of black strikers established a permanent community, building "their own shacks, often-time using abandoned or waste building materials" after being ejected from company housing. "Oldtimers around Dora still call it Uniontown," recollected a white UMW retiree in 1977, "because black union men . . . went in there and built their own crude houses so that they would never again have to live in company houses that they could be thrown out of." Marvel management testified to the militancy of black strikers before the Kilby commission, complaining that "in the negro camp, to a man, not a single negro has gone back to work. They just sat there in the negro quarters and held that camp." Moreover, many white miners freely acknowledged that black unionists bore the brunt of repression in the coal camps. Even if most did not consciously reject racism, their experience fighting alongside blacks could not help but impress upon whites the fact that staunch black unionists paid a high price for their loyalty to the UMW and were therefore deserving of respect and solidarity.[53]

It is difficult to arrive at any conclusion other than that the military, along with private and public law enforcement authorities, singled out black strikers for especially harsh treatment. White miners were by no means immune to rough treatment from the military. The complaint from a Brookside UMW member that "we have not got freedom as people but [live] in bondage like mules" could as easily have been written by a white miner as a black. The Pratt City local, for instance, held their meetings under the shadow of a machine gun mounted "on the platform of the meeting hall," and miners there complained that they were being "hounded and intimidated by troops as though they were criminals." Officials at Marvel forwarded a request that the union arrange medical treatment for strikers: "Judging from the way the State Guard is acting up around here," a union official predicted, quite seriously, "it will not be long before some of us will be ready for the Hospital or Semitary [sic]."[54]

Throughout the district, however, authorities meted out repression disproportionately against African-American strikers. Several weeks into the strike the body of the black UMW miner Henry Junius was found buried near Roebuck.[55] In December twenty soldiers "emptied a machine gun placed . . . at the heart of Pratt City . . . into an embankment . . . on the side where the colored people live" and then proceeded to terrorize the patrons of the small black business district there. At Boothton mine bosses "caught beat up and run [off] a negro named Joe Dee" who had recently joined the union and "told one of the old negroes named Will Parker that it was moving day and if they had to come to him again [they] would use lead." At the Red Feather mine at Blocton, company gunmen, escorted by "a marshal by the name of Chew went to several negro houses with guns and pick handles and threatened to kill them if they did not appear in the morning for work." At Palos the commander of the State Guard made things "very unpleasant," "using profane language, cursing and abusing [strikers] and threaten[ing] the colored miners that 'if they did not go to work he would run them in the Warrior River.'"[56]

Common suffering naturally generated a degree of camaraderie between black and white unionists and laid the basis for a challenge to racial protocol. Though it concealed a primary concern with their ability to turn a profit, the operators' harping upon the UMW's defiance of local custom confirms not only the salience of race in the early twentieth-century South but the uniqueness of the interracial ethos emanating from the coalfields. Chief among the operators' objections to the union was the contention that it had been the UMW's "ambition and purpose . . . for years to organize [the] negroes and through them to regulate and control Alabama's coal production." The "representative Citizen's Committee" closely aligned to the mine owners pointed out, disparagingly, that within the UMW "negroes have been elected to high office . . . and their authority and counsel are respected and obeyed." And among the evidence submitted against the union during hearings held before the Kilby strike commission was an affidavit charging that the black UMW vice president Joe Sorsby had been observed "very often [dictating] to the white stenographers with his hat on his head, and with a cigar in his mouth." Scandalously, in the operators' view, white UMW members had been overheard "address[ing] the negro as 'Mr. Sorsby.'"[57]

Exceptional though the UMW's interracialism was by contemporary standards, however, among most white miners—including union stalwarts—thoroughgoing egalitarianism never managed to supplant the racialist assumptions they shared with every stratum of the white South. Joint struggles around shared material interests—none of which were more dramatic or protracted than the 1920 strike—provided a powerful incubus for large-scale questioning of such assumptions. Certainly nowhere in the early twentieth-century South were the traditions of racial protocol challenged more forcefully than in the Alabama coalfields. But coal unionism was hard-pressed to survive even as a purely bread-and-butter affair, let alone emerge victorious in a confrontation with its adversaries or—even more ambitious—detonate the wider, more general working-class rebellion that alone could provide breathing space for interracialism to take root. The unfavorable circumstances peculiar to the South, intensified by the belligerence of capital and the pro-business posture of government in postwar America, blunted any potential for the emergence of a durable egalitarianism.[58]

As an interracial organization, the UMW became a site where black miners attempted to push the limits of racial justice, but in a society that inscribed white supremacy so indelibly in everyday life, relations between black and white trade unionists inevitably reflected both the potential for a break with tradition and the enduring power of racial custom to shape the present. The most benign example of this can be found in the separate supply requests (categorized by race) that local unions forwarded to UMW headquarters throughout the strike: in December the Straven and Garnsey union locals submitted separate lists for Christmas toys for black and white children in the camp.[59]

More commonly, however, black miners and their wives expressed resentment at the disparity in the union's treatment of black and white strikers. They complained at being cheated out of supplies by white local union officials. A "Colord Union Woman" penned a remarkable letter to UMW officials in early December in which she expressed her pride that the "colord miners of Marvel [on strike since June] have never broken ranks yet" and her resentment that, of the funds sent to local officers to buy shoes, "very few of the colord got any thing at all." "It is not fair," she protested. "If we are to strike togather [sic] I think we should sheare [sic] the same. . . . If something is not done . . . for these peoples I think that the strike will soon be over at this place." The UMW organizer R. W. White reported from West Blocton that "out of the eighty some pair" of shoes sent by the union, blacks "only got thirteen pair." "There is some of the colored people there that is naked and bare footed," he wrote, "but they will waite untill [*sic*] they can be supplied." D. J. DeJarnette, who only weeks earlier had offered to tour the Black Belt in an attempt to check the operators' strikebreaking plans, wrote Bittner in mid-December to complain that the Brookside local was "not treating [him] right at all." "With nothing to eat for three weeks I tufted it out hungry and said it was our fight . . . there [is] not one drop of scabbing blood in me but now I sure enough barefooted . . . and I am naked to with my last pants on and they are all torned [*sic*]." DeJarnette's condition was so extreme that he could not even leave the house to post his letter but had to entrust it to his daughter, whose clothing was as yet adequate enough to allow her to be seen in public.[60]

Although it does not by any means negate the black mining community's grievances over racial discrimination within the UMW, the backdrop to their complaints was a situation of increasing general deprivation throughout the district. The president of the Straven local wrote Bittner in early November for advice on "how to give [relief supplies] out, for there is not enough to give to all the ones that are in great need." Only days earlier Bittner had received a letter from national headquarters informing him confidentially "that the surplus of funds . . . [is] practically gone." Nor did the inequitable distribution of union rations necessarily emanate from racial divisions: from Dora came the complaint that "the men that has control of the splies [*sic*] out here just gobbles them up and some of them gets a plenty and some gits none." Conditions were such at Coal City that Mrs. Lennie Creel wrote to complain that although women there had faithfully implemented a directive to organize a sewing club more than five weeks earlier, they had "never received anything to sew." "Some of the people are getting impatient," she wrote, adding that "you can't blame them much." The miner Henry Woodford wrote from Cardiff that he had "to keep my children home from school for the want of shoes . . . and other children [are] laughing because my children cannot get Union shoes." At Marvel local officials gave up on getting shoes and requested that the union send "a few pounds of shoe leather and some tax [*sic*]." And at Kimberley parents faced the prospect of being sent to jail under the compulsory school laws because they could no longer afford to pay a teacher.[61]

The conduct of the strike suggests that the UMW's reputation for egalitarianism is subject to several qualifications. First, day-to-day relations within the union reflected and were profoundly shaped by the racial customs extant throughout southern society. Moreover, even at its height the bearers of the interracial tradition never felt confident enough or compelled for whatever reason to pose a frontal challenge to Jim Crow. What stands out in the 1920 strike is the way in which working-class interracialism navigated the treacherous boundary between conforming to racial protocol and challenging it openly. Contrived and implemented under a near constant state of siege, this policy derived in part from the racial outlook of white miners and union officials. But it was shaped more profoundly by the volatility injected into every dispute by the operators and their allies. The union's guarded approach to the organization of women's auxiliaries in 1920 illustrates this dynamic. One of the indelible memories of the 1908 strike was the press's condemnation of the UMW's organization of interracial women's clubs to support the strikers. "Think of it," an *Age-Herald* columnist had remarked at the time. "White women and black women meeting on the basis of 'social equality' indeed." The racial defensiveness that characterized the UMW effort twelve years later can be clearly seen in the leadership's explicit instructions that "the wives and daughters of the white miners [should be] organized in one body and the wives and daughters of the colored miners in another."[62]

Further evidence of the union's hesitation to openly defy racial custom can be found in its tortured attempts to dodge the "social equality" bugaboo. Under attack for advocating "social equality" between blacks and whites, District 20 officials attempted to deflect the charge by publicly framing the issue in class rather than racial terms. While this was in some sense a conservative strategy, such an approach did not necessarily preclude militancy, and union officials proved themselves adept at turning the operators' logic on its head.

"These colored men are in the mines because the steel corporations . . . put them there, and they had to go to make a living, the same as you white fellows," Van Bittner told a mixed assembly of miners in October. "There is nobody working in these mines for their health. . . . I don't care what a man's color is so long as his industrial relations . . . is all right—so long as he is a good union man. [We cannot] fight this issue on the question of race, and discriminate against the colored man. . . . We are going to get the same wages and the same conditions, industrially, and the same living conditions for these men in these mining camps . . . as we do for the white men. (Great applause)." Addressing the specter of "social equality," Bittner assured an audience of railway workers at Albany in early November that the union was "not asking that the wives of the white coal miners be put on social equality with the wives of the white Coal Operators." "Certainly," he said, "we have taken [the black miners] into the Union, and we are going to keep them in the Union, because we may as well recognize . . . that the wage . . . the negro worker gets is going to be the wage that the white workmen are forced to take . . . it is only one race question involved and that is the race between the UMW and the ACOA."[63]

The union's strategic calculation that advocating industrial rather than social equality would defuse the operators' race-baiting was proven wrong in the course of the 1920 strike. The mere presence of blacks and whites in the same organization ensured that, regardless of its formal policy, the union would be subjected to charges of racial subversion. One of the intriguing questions raised by the conduct of the strike is whether a more frank and open defense of interracialism, supplemented with a broad internal campaign to prepare white members for the vilification that would be forthcoming from the operators and their allies, would have been more successful. The much larger question is whether the UMW could have prevailed in 1920 Alabama under any circumstances.

The union's strategic concentration on shutting down coal production with mass meetings and militant pickets seems in hindsight to have been both appropriate and necessary. But this strategy derived in part from an assessment that their disadvantages on every other front were insurmountable. Bittner's complaint that "every force of government and everybody else not directly connected with the trade union movement is against us" was confirmed by a number of other observers. Mary McDowell, dispatched to investigate conditions in the district by the Interfaith Commission on the Church and Social Service, reported to her directors in Chicago that "the press and the public seem overpowered by the corporations. . . . Alabama . . . is so backward and its public opinion so in the grip of the TCI," she wrote, "that it is said that the Gary group [U.S. Steel] own and control the land, government, etc." Like the miners, McDowell considered Governor Kilby "an operators' man"[64] and wrote that the employers' influence extended even to religion: "The Methodist ministers are no use in Birmingham and only one Baptist seemed unafraid, a Mr. Durant, who has a church in a workers' community."[65]

With such an impressive array of power at their disposal, leading coal operators felt little urge to settle for a compromise with the union and instead held out for a complete victory as in 1908. When in the first weeks of the strike Governor Kilby appointed a three-man commission[66] to make recommendations for settling the dispute, operators circulated the somewhat contradictory statement that while they "pledge[d] themselves to cooperate with the commission," they "[would] not recognize the union nor deal with it, directly or otherwise." It was their pressure that had brought troops into the coalfields after Leon Adler's death, and the mine owners continued to exert tremendous influence over how the military was deployed throughout the strike. The federal mediator William C. Liller attempted on a number of occasions to broker a compromise between the ACOA and the union but found his efforts constantly undermined by a core of large-scale operators. In correspondence with the governor, Liller revealed that "a large number of Operators desire to end the strike, but the [ACOA] Operator's Committee, which dominates the situation, prevents them from doing so." These leading operators were "determined . . . to fight it out now to the bitter end and 'wipe out the last vestiges of unionism in Alabama' and . . . if necessary they will keep on the fighting if it takes every dollar they have at stake." Not surprisingly, Liller eventually found himself in disfavor with the ACOA, who brought pressure upon local politicians and succeeded in having him removed from Alabama.[67]

Confronted with such awesome power, the UMW had little choice but to mobilize the full potential of the mining community and its supporters in the local labor movement to fight it out on the picket lines. Despite some friction between the UMW and the local Trades Council, organized labor sponsored a number of successful mass meetings in the district. Angered by partisan coverage of their struggle in Birmingham's daily newspapers, strikers turned avidly to the *Labor Advocate* and the *Advance* for strike news. "We wear the print off [the *Advance*] when we get passin' it from one to the other," a Roebuck miner reported. "The operators can howl because the *Advance* won't lie for them," another wrote. "The *Birmingham News* is the Romans [of our age], the *Age Herald* [the] Centurions. And the *Advance* repsent [sic] Christ."[68]

Complementing the strikers' organization into picketing committees, district officials recognized that coal camp women would play a pivotal role and set about organizing female auxiliaries. The initial motivations for organizing women were purely pragmatic and had little to do with challenging their traditional roles. Part of the function of the women's auxiliaries was the organization of sewing clubs, the importance of which grew as the strike wore on and strikers' families reported being literally without clothes. But such organizations obviously reinforced rather than challenged women's domestic roles in the camps. Nevertheless, mining women seem to have valued the limited space the auxiliaries presented for asserting themselves. At Blocton, the "Ladies' Union Club" claimed sixty-five white and over a hundred black women. Carrie E. Goode of the Brookside local toured the coalfields reminding women's auxiliaries that "we haven't gathered here to spend our time uselessly. . . . We want our men to put over one of the greatest victories ever known in strike history, and we should feel that it is just as much our fight as it is theirs."[69]

UMW officials were astute enough to realize that as hardships set in and began to breed disaffection among strikers and their families, operators would look upon those women who had not been won to solid support of the union as a potential wedge with which to weaken the strike. Miners' letters reveal that the strike did generate increased domestic tensions, though not always because women constituted the weak link in UMW militancy. No doubt a number of miners could relate to the predicament that Jobie Jones found himself in at home. Jones reported that he and his wife "got into some difficulty [when] she wanted [him] to scab and I would not do that." Jones ended up assaulting his wife, after which she called for a company guard, forcing him to leave the camp to avoid arrest. Alice Moore wrote Bittner from Adamsville to warn that if he didn't resolve the strike soon she was "going to make my man go to work."[70]

Elsewhere, however, women outdid their husbands at militancy. Oscar Fay, a local official at Parrish, vowed that he would give the union his "best service because I have nothing else to do but live for my fellow worker and my wife and baby." Fay attributed his support for the union to two factors: "One it is in my blood and can't be worked out," he asserted, "and another is I love my wife and baby and if I should scab [they] would leave so you see my fix." A Burnwell miner praised his "little wife who tried to con cold [*sic*] me when the company gunmen was a resting [*sic*] me and pinching me on every side on a count [*sic*] of the union." "She . . . have never said to me let the union go [even though] her everyday clothes and under wear is about gone." And union women showed themselves equal to their menfolk in drawing sophisticated political conclusions from their experience during the strike. After a Birmingham "lady" penned a letter to the *Age-Herald* reminding miners that their homes were their "castles," a striker's wife responded bitterly that the author "seems to forget that even our castles have been shot into by our noble soldiers . . . all because we have raised up and shown a spirit of resentment against such merciless principles shown by the [operators]." "We are going to get rid of these blood-suckers and put them to work," she vowed defiantly, "and not let them live off of the toiling masses."[71]

What seems most remarkable about the strike in retrospect is not that it eventually went down in defeat but that miners and their families managed to hold out for so long against such formidable odds and under such heavy repression. Despite the ability of the large operators to sustain adequate levels of production, they became increasingly frustrated with the UMW's unwillingness to cave in. When in mid-December the union's special convention seemed to momentarily revive strikers' flagging morale, their adversaries—Governor Kilby, the commander of the Alabama National Guard, R. E. Steiner, and the ACOA's Combat Committee—determined that the time had come to put the strike down forcefully. On December 23, 1920, Steiner issued orders forbidding the holding of regular business meetings of the UMW locals. Kilby dispatched an undercover squad from the Alabama Law Enforcement Bureau to Birmingham, where they shadowed and harassed UMW officials and sympathizers, committing a brutal assault against the editor of the *Advance*, Phil Painter. Van Bittner received warnings that he would be lynched or otherwise removed from the scene: "All eyes seem to be centered on you," agent Liller informed him confidentially. "I have good reasons for believing . . . that [the operators] intend to 'pull' something on YOU and I am writing . . . so that you may be prepared to meet ANY situation that might arise."[72]

The situation took a marked turn for the worse after the New Year. On the second of January Steiner related to Kilby his belief that "this week will prove the turning point in trouble. While the strike is broken in that coal production is normal . . . yet the strikers are not broken and it is this last situation we have to deal with." Conrad Austin, the head of the state's undercover squad, would later testify that Steiner had told him of Kilby's dissatisfaction with their progress. "The Governor feels like this strike ought to be brought to an end and these Agitators ought to be run out of here. . . . He feels like we ought to whip hell out of them," Steiner was said to have reported. Liller protested at the same time that "a representative of one of the big Corporations here" had told him that "the only way to get anywhere and put an end to the [strike] is to start 'pulling the rough stuff.'"[73]

The effects of this renewed effort to break the strike became evident almost immediately throughout the district. "The public . . . are being treated very cruel by the soldiers especially the negro race," an anonymous correspondent from Pratt City informed Kilby on January 9. "The soldiers are knocking thim around and kicking thim around with their rifles and pistols and will hardly let thim stand at the trolley car stations long enough to catch a car." Guardsmen claimed to "have orders to put the men to work or put them in jail." At Marvel soldiers "rode all over the quarters . . . and every where they found three or more men sitting on a porch they ordered them to get in the house." Eleven black strikers caught "in an empty house playing a fun game of dominoes" were arrested by troops at Marvel on January 5, and a white miner there was bundled into a military vehicle, whisked away to a remote location, and beaten. Throughout the district came reports of a disturbing new development: Ku Klux Klan nightriders had "set out upon a reign of terror against the miners, . . . parad[ing] through parts of the mining

MILITARY ORDERS

ISSUED BY

BRIGADIER-GEN'L STEINER

Under and pursuant to the Constitution and Laws of Alabama and the General Laws covering the present situation:

1. I have assumed jurisdiction of all mines and territories adjacent thereto throughout the Mineral District.

2. Marching on the roads will be prevented.

3. All persons bearing arms, not authorized to do so by law, will be disarmed.

4. Mass Meetings and Assemblies in the open will be dispersed.

5. Addresses to Mass Meetings and Assemblies in the open will be prevented.

6. Regular business meetings at regular places of meetings will be permitted, provided notice to troops is given, but no incendiary or inflammatory speeches will be permitted.

7. Intimidation, threats and all acts tending to provoke a breach of the peace will be prevented.

8. Protection will be afforded to all law-abiding persons and no sides will be taken except against the law breaker.

General Steiner's declaration of martial law during the 1920 strike was posted throughout the mining district and its regulations stringently enforced, to the point where UMW supporters complained that church and fraternal meetings were banned. Note especially Steiner's ban on mass meetings, which dealt a crippling blow to UMW attempts to halt production in the district. (Kilby Administration Files, SG 22121, folder 21; courtesy of the Alabama Department of Archives and History, Montgomery, Ala.)

field at the dead hour of night, for the purpose of frightening the union miners into deserting the union."[74]

The most serious incident of the strike, and one that appears to have arisen directly out of the new campaign to crush the union, occurred in Walker County on January 5. Ten days earlier Adrian Northcutt, a union miner and itinerant minister, had been summoned from his home in Nauvoo by members of Company M of the Alabama Guard. Soon after he had departed with the soldiers, his family heard seven shots "fired in rapid succession" some distance from the house, and his twenty-two-year-old son-in-law, Willie Baird, rushed to the scene to find a soldier standing over the miner's lifeless body. An argument developed between Baird and Private James Morris that ended with Baird shooting the soldier and fleeing into nearby woods. Turning himself in to authorities three days later, Baird was taken to the Walker County Jail. On January 5 nine guardsmen, members of Company M, entered the jail accompanied by the brother of a mine superintendent, subdued the sheriff on duty, and removed Baird, lynching him and filling his body with bullets.[75]

Reaction to the Baird lynching illustrates how deeply polarized the mining district had become during the course of the strike. To the UMW the outrage "demonstrate[d] that the present military and law enforcement authorities . . . are either directly or indirectly responsible for creating a reign of terror" that had "deprived the miners of every right guaranteed them by the laws of this State and the Constitution of the United States." The operators and their allies, however, launched a fundraising campaign to defray the costs of the soldiers' legal defense. The former governor, Braxton Bragg Comer, who had earned the enmity of miners during the strike twelve years earlier, justified his own contribution to the fund with the assertion that the killing had "some element of self-defense in it." The Birmingham attorney Horace Wilkinson, shaken by the events, protested that "for the first time in the history of the state some of the substantial citizenry . . . seem to be openly endorsing mob law." He condemned the soldiers' defense team for conspicuously summoning a "negro official of the mine workers' union" (vice president Joe Sorsby) whom it had no intention to use, "but was kept at court, and pointed out by friends of the accused as a union leader to PREJUDICE THE STATE'S CASE."[76]

Reeling under the barrage of overwhelming repression, the UMW was dealt another heavy blow when toward the end of January the Alabama Supreme Court upheld lawsuits brought by the operators to eject strikers from company housing. Strikers had attempted, where possible, to continue to occupy the coal camps as a way to block the importation of scabs and to assure themselves of shelter during the strike. Liller believed that the ejection writs were "only for vindictive purposes and to force the miners to return to work," and once again the UMW attempted to adapt and persevere. They secured from the national organization thousands of tents for their membership, a form of accommodation that some miners seemed not to mind. "Most of us work hard for . . . just a little shanty . . . but those places is all we have out of our whole life earning and we value them as highly as those operators does their millions," one miner wrote, along with a request that he be allowed to keep the tent furnished by the union, finding it "so much more comfortable than those barns we moved out of."[77]

Notwithstanding the determination displayed by their members, UMW officials realized by the end of January that the strike could not continue. Despite having expended a fantastic sum (fifty-thousand dollars per week over a five-month period), they seemed no closer to wresting union recognition from the operators at the end of January than they had been at the outset of the strike in September. Bittner's hope that government intervention could produce a compromise that would at least legitimate the organization in Alabama was dashed when Liller was ejected from the field in early February. By the middle of that month, "fatigue had set in and the strike began to weaken. Men began drifting back to work [while] the operators held steadfast to their position of non-recognition . . . rising production figures plus the widespread intimidation from operators and the state government had taken their toll on the miners' morale." On February 22 Bittner agreed to submit the strike to Governor Kilby for arbitration, apparently convinced that the governor would, in the interest of securing industrial peace, go some way toward meeting the union's demands.[78]

On March 19, 1921—more than six months after the strike had begun—Kilby issued his decision, ruling against the UMW on all counts. He condemned the strike as "illegal and immoral" and recapitulated practically verbatim the operators' claim that its effectiveness had been due to the high proportion of gullible, "easily misled . . . southern negroes" in the union's ranks. Kilby seemed determined not only to seal the operators' triumph but to finish the rout of the UMW. He insisted that, having brought the men out on strike "without just cause or for the purpose of remedying any grievance," the union should be held responsible for feeding and housing miners and their families until steady work again became available. He made no effort to compel the operators to rehire those who had struck but instead called upon the employers to demonstrate their "graciousness" by "re-employ[ing] [those] who struck as . . . places may be found."[79]

Unmoved by appeals from starving miners and their families that he compel the operators to reemploy those out of work, Kilby called upon the union's enemies in Birmingham's business community to undertake the miners' material sustenance as a matter of philanthropy. Few protagonists on either side could have discerned much ambiguity in Kilby's appointment of the Birmingham banker O. T. Smith, who had been one of those clamoring loudly for vigilantism only months earlier, to oversee the "charity work." "The miners, who have been contending for what they believed to be their right, have put up with a great deal more than they should have," the editors at the *Labor Advocate* contended. "Now it seems there are a lot of people who want to add insult to injury." Despite a massive commitment to organizing the Birmingham district on the part of the national UMW and a colossal, heroic effort on the part of rank-and-file miners, the union had been unable to match the power of Alabama operators and their allies. The vice president of the Alabama Fuel and Iron Company told his stockholders that while "this has been a hard fought battle . . . the hardest fight which has ever been waged in Alabama," he considered their efforts worthwhile. "Notwithstanding the money expended by our company [over sixty-three thousand dollars for AFICo alone, he estimated], it has been cheap, as we have won a complete victory, and . . . it will be many years before we are annoyed again by the UMW and their agitators." The coming years would bear grim testament to the accuracy of his words.[80]

* * *

> "Go now ye rich men weep and howl for your miseries shall come upon you . . . ye have reaped treasure for the last day . . . ye have lived in pleasure on the earth and been wanton; ye have nourished your hearts as in a day of slaughter."
> —James 5:1–5

More than a half-year of bitter, frequently violent confrontation had left Alabama's mineral district deeply polarized. One casualty of the upheaval was the relationship between white miners and the Democratic politicians who had overseen the crushing of the strike. Shortly before the New Year, the seventy-six-year-old Confederate veteran J. F. Boston, who had worked the mines since before the turn of the century, wrote Kilby to warn him of the danger that miners would desert the party wholesale. Noting that UMW members at Adger were "as law abiding men as is in the state," Boston warned that "the next election will go republican as men are getting tired of democratic rule." Well before the announcement of Kilby's decision, miners expressed anger not only at their treatment by the military but at the blatant hostility of authorities at every level. In this charged atmosphere, the subterranean class politics present but infrequently expressed throughout the mineral district became much more pronounced and overt. A UMW member wrote the *Labor Advocate* to urge that labor stand its own candidates. "If we don't take a hand in the reins of government," he warned, "it is going to run so far away from us we will have to refer to some library to find out whether we belong to the United States or some other country." The strike seemed to confirm, in his view, that "the laws . . . are only gotten up by a few of the monied class to keep labor back." A Blocton miner reasoned that "this so-call Dimircrat and Ristercrat Democracy" wanted to "git the min back to work so there Big Money will go to rooling . . . again." "People here has come to a pass," he vowed. "They had just as soon dye as to live under such ruling as the Democrats is imposing. . . . This strike must not be lost for if it is god pitie the pore class in Alabama."[81]

UMW officials had submitted their claims to Kilby from a position of weakness and obviously did not expect to be fully vindicated by his ruling. But they hadn't expected such thorough humiliation either. The decision left them deeply embittered and outraged, so much so that they contemplated attempting to revive the strike. Bittner toured the coalfields in the days after Kilby's announcement urging miners to prepare for a return to confrontation. "There has not been any strike yet," he told a group of miners at Blocton on March 30. "We are just going to start. . . . If there is any scabs left here in Blocton thirty days from now, you men ought to be chased out of the state. . . . It has been entirely too healthy for scabs around Blocton. . . . You are between the devil and the deep blue sea; if you don't keep the union here you are going to starve to death."[82]

That Bittner's rhetoric reflected the real sentiments of many striking miners and their families is indisputable. A deluge of reports and letters received by Kilby's office in the days after the strike's termination confirmed the deep anger. Mrs. T. E. Reeves of Aubrey wrote Kilby to apprise him of the "real spirit of the coal miners." Van Bittner's speeches "and letters of I. W. W. or socialism has set them a' fire," she warned. "One man in this camp said that every coal tipple in Alabama should be burned, they could form a company and turn them all in one night before you could get troops here." "Every miner takes the *Advance*," Reeves observed, "and its the most unamerican paper I ever heard of. . . . Very few of them have enough money to purchase one weeks' supply of food. . . . I trust you will rid our state of this dangerous man [Bittner] and paper before [the miners] draw their last food."[83]

Hyperbole aside, Reeves's assessment of the volatility present throughout the mineral district seems accurate. Desperation stalked the coalfields, and letters flooded in to Kilby and UMW officials from miners and their wives who were literally on the brink of starvation. A sentiment for vengeance—against Kilby, against the operators, against the powerful forces who had broken the strike—seems to have been almost universally embraced. From the UMW local at Parrish came a letter informing Kilby that his "decision leaves . . . four hundred and fifty-two men, women and children simply 'at sea.' No home. No shelter. No food. No clothing. No jobs. No nothing. . . . There is not a red-blooded American . . . that is going to set down with his hands folded and see the very life blood sapped out of his wife and babys in this manner." From Wylam miners wrote that it was "indeed a humiliating task for us to make an appeal to you [Kilby] for assistance . . . nevertheless we have no apologies to make whatever. . . . We will not be responsible for any conditions that may arise from this unhappy situation. You have spent thousands of the people's money to protect the coal operators." Perhaps the sharpest threat came from a miner's wife at Blocton. "We all look to you," she warned Kilby. "You and the operators are held direct[ly] responsible for these men. . . ." "This is our state," she concluded. "We will fight for freedom. There will be a revelution [sic] start right in Alabama [and] it won't be weeks about it either."[84]

Given the capacity shown by UMW members and their families for militancy and self-sacrifice over the duration of the strike, it would be a mistake to dismiss such protests as the idle threats of the defeated. No doubt many would have preferred to remain out rather than return under the humiliating conditions dictated by Kilby. But in reality the strike had been broken, and broken decisively. The union was shattered, and it would be more than a decade and a half before coal miners regrouped—emboldened by the wider upsurge propelling the organization of the CIO—to challenge Birmingham district operators once again.

In the meantime mine owners went about resetting the foundations of the authority they had enjoyed before the war. By June Kilby was receiving complaints from former strikebreakers at the Winona Coal Company that their employer was cutting wages. They asked that he "withhold" the petition from "officials of the company . . . for they will recognize the hand and the writer will be discharged." Within six months operators throughout the district had instituted a "substantial reduction in wages," a measure that the ACOA chief James Bonnyman arrogantly contended was adopted "at the request of the men." "Uncle Charlie" DeBardeleben expressed his pleasure that productivity was returning to normal levels now that "labor . . . was getting over their exaggerated ideas contracted during the War." When the ACOA met later that year for their annual meeting, the general counsel for

LOOK LOOK

Wanted 1,000 Colared miners with families and 500 single colored coal miners within the next thirty days for West Va. Virginia and Kentucky.

Come Prepared to go

What wages were you getting 2 years ago? What are you getting now? If you are satisfied, stay here, if not see Jones Labor Agent.

Ships every day on every train. If you cant go yourself, come and see your friends go.

L. W. JONES Labor Agent

12th. Ave & 19th. St. Bessemer Ala.

Licensed to ship anywhere in the United States.

Alabama operators were infuriated by "outside" attempts to entice their mine labor northward but frequently engaged in the same methods. Posters like this appeared throughout the district in the aftermath of the UMW's defeat in 1920. (DeBardeleben Coal Company Records; courtesy of the Birmingham Public Library Archives, Birmingham, Ala.)

the National Manufacturers' Association, Judge James A. Emory, delivered the keynote address, entertaining the gathering with a timely talk on "The Open Shop Movement—Its Justification and Progress." And in a fitting epitaph to the convulsions that had rocked the district only months earlier, operators were entertained on that occasion by the DeBardeleben Coal Company's Sipsey Colored Brass Band. Under such conditions the city of Birmingham celebrated its fiftieth anniversary. "This is the Magic City of the World," sponsors of the festivities boasted, "the Marvel of the South, the Miracle of the Continent, the Dream of the Hemisphere, the Vision of all Mankind." Those upon whose labor the miracle had been built were conspicuous by their silence, their capacity to object having been crushed in a bitter confrontation that the dinner guests would prefer to put behind them.[85]

Conclusion:
Bringing the Employers Back In

SURVEYING THE CARNAGE that had been visited upon Atlanta, Georgia, in the days following that city's gruesome 1906 race riot, the liberal northern journalist Ray Stannard Baker articulated what he believed to be an axiomatic explanation of race antagonism in the American South. Praising the "strong men of the city" who, in defense of their "pride, their sense of law and order, [and] their business interests," had responded energetically to the Atlanta crisis, Baker attributed the sudden collapse of racial détente to the fact that "the poor white hates the Negro, and the Negro dislikes the poor white. It is in these lower strata of society, where the races rub together in unclean streets, that the fire [of race hatred] is generated." "The ignorant Negro and the uneducated white," he explained to readers of *Survey* magazine, "There lies the trouble!"[1]

Baker's postmortem on the events of 1906 did not survive the scrutiny of later generations. Much of the responsibility for the Atlanta pogrom came eventually to rest upon the state's most respected newspapermen, who had contributed mightily to the public hysteria, fueling the city's lynch-mob atmosphere. Governor Hoke Smith, credited in Baker's account with having provided refuge for black Atlantans during the peak of the violence, had only been recently installed at the state capitol after an electoral campaign artfully constructed around the pledge to strip black Georgians of the franchise; he won a reputation in later years as one of the state's most fervent and unscrupulous racial demagogues. Only four years after the rioting, in the midst of a series of "hate strikes" aimed at dislodging black Georgians from skilled labor, the National Negro Conference, meeting under the auspices of the NAACP, passed a resolution refuting the notion that "such movements [were] spontaneous and natural." The "blame" for such outbreaks "should be placed . . . not on the illiterate and ill-informed mass of poor whites," delegates insisted, but upon "men like Hoke Smith [and like-minded] leaders who make it a profession and life work to foster race hatred." It was white elites, they argued, and not the white rabble who "bear away the fruits of each successful crime against humanity but place responsibility for . . . excesses on their ignorant followers."[2]

Although Baker's explanation of the Atlanta Riots has failed the test of time, the more general assumptions underlying his account have displayed a remarkable longevity in popular and scholarly comprehension of America's "race problem." The paternalist interpretation of southern history, which depicts an embattled, progressive white elite standing guard between warring camps of poor whites and blacks, continues to exert a powerful influence upon public understanding of the region's past. The time-honored view "that wealthy, cultured whites, including the large ex-slaveholders, cherished a kindlier feeling toward Negroes and exhibited less race hatred than their lower-class brethren," labeled by the historian Allen Trelease a "self-gratifying myth," has been absorbed wholesale into liberal thinking about race.

The classic expression of Baker's thesis was advanced in the 1940s by the Swedish sociologist Gunnar Myrdal. In his *American Dilemma*, Myrdal explicitly rejected the formula that had been advanced by southern radicals during the Depression, which linked racial progress with a wider challenge to the power of the southern ruling class—one that would raise the position of white workers along with that of the mass of blacks. As one perceptive critic has noted, Myrdal's argument "broke with much of the scholarship of the 1930s by placing its stress on the competition and hostility between working-class whites and blacks rather than on their basic compatibility of interests." In Myrdal's estimation, "'lower class whites ha[d] been the popular strength' behind anti-black public policies," and their "hatred . . . toward Negroes" was the "foundation upon which the appeal of white supremacy rested." It followed logically that "the more natural ally for blacks was the 'upper class of white people,'" regarded by Myrdal as a "non-competing group." His prescription for the problem of race hatred, like Baker's, was patrician guidance-from-above and moral enlightenment for low whites.[3]

This brief digression from the northern Alabama story is worthwhile not only because Myrdal's analysis exerted a strong pull upon a generation of (now nearly extinct) American race liberals but because the liberal perspective has worked its way, insidiously, into contemporary radical scholarship on the tangled relationship between race and class. Although they would strenuously deny such intellectual lineage, a number of prominent left-wing scholars have repackaged the paternalist fallacy and today wield it as a positive advance over earlier, marxist-influenced historiography.

It is worth emphasizing what is not being disputed in this ongoing debate. Antiblack racism has been, to borrow Marx's phrase concerning English workers' prejudice against the Irish, the "secret of the impotence" of the American working class. The most obvious problem with the current emphasis on the psychological "wages of whiteness" is not its assertion that those at the bottom of white society often accepted black racial inferiority as axiomatic or that they were capable of perpetrating brutal racial atrocities in defense of white supremacy. Antiblack prejudice certainly permeated every stratum of the white South during this period. White workers absorbed the racial chauvinism then intoxicating the region as naturally as did better-off men and women of their race and probably figured more prominently in racial attacks than those from the upper ranks of white society. But to conclude from this, as Baker did, that the enmity between poor whites and blacks suffices as an explanation for racial violence—rather than as a register of its pervasiveness—only confounds the problem of unearthing the roots of racial animosity in the New South. Similarly, the assertion that racism had deep roots in the collective psyche of white workers does not begin to explain the variety of outcomes labor historians are beginning to uncover in cross-racial interaction at the bottom of southern society.

This study attempts to illustrate that in whatever ways pragmatic interracial collaboration stopped short of full equality in the Alabama UMW, by any measure the main beneficiaries of and the principal force in maintaining black oppression in the Birmingham district were its major steel, iron, and coal employers. That assessment can be extended with only slight modification to fit the New South as a whole. The terms upon which black and white workers came together in the early twentieth-century South were set by white elites. Black miners seem to have understood this dynamic profoundly, despite concerted, well-financed attempts to prevent them from reaching such conclusions. And, at their best, white miners seem to have grasped that the degradation of black labor was in some way aimed at them as well. This did not make them paragons of racial virtue, nor did it inoculate them against the potent contagion of race prejudice. But the developments do raise a profound question about how, in the process of defending their material interests alongside black southerners, even a small number of white workers questioned the ideological orthodoxy that faced them at every turn during what Leon Litwack calls "the most violent and repressive period in the history of the United States." The answer lies, I believe, in the singular potential for joint struggles around material interests to begin to break down racial divisions.[4]

A proper focus on the historical context in which black and white workers interacted is essential not only for understanding the high points of interracial collaboration but also for plumbing the import and essence of the South's most intense racial confrontations. White employers did not generally open employment opportunities to southern blacks out of a sense of racial benevolence. The strategy pursued by Birmingham operators after the turn of the century—displacing white labor with lower-paid (and disfranchised) black labor—was enthusiastically engaged in by their counterparts elsewhere in key sectors of the southern economy. The notorious Georgia "Race Strike" of 1909 began when Atlanta railroad executives began "replacing white firemen with blacks at lower pay" and "without notice fired ten white assistant hostlers and replaced them with Negroes at fifty cents per day less in wages." Significantly, in their early demands white strikers asked "that the white hostlers be reinstated [and protested] the use of black workers to keep white wages down [but] did not ask that recently hired black firemen be removed, that the railroad cease to hire black firemen, that a quota of blacks be established, or that all black firemen be replaced." As the strike wore on, however, and as management dug their heels in, racial antipathy (egged on by northern union officials, incidentally) came eventually to dominate the strikers' outlook so that by the strike's end "the most ominous idea running through poststrike statements was that certain kinds of work should be closed to Negroes."[5]

The apparent ease with which an industrial dispute was transformed almost reflexively into a racial confrontation would feature prominently in any faithful reconstruction of these events. The pervasiveness of white supremacy and the traditional hostility of the railroad brotherhoods against black workers practically guaranteed such an outcome. But this would tell only half the story and would relieve company executives from any responsibility in pursuing what amounted to a calculated strategy of racial provocation. The same could be said of the Birmingham steel companies' successful attempts to bar unionism from district mills in 1918 and 1919 or of the policy pursued in the southern lumber industry or on the docks of New Orleans. White workers may indeed have carried the contagion of racism deep within their psyche, but it was the continual and systematic pitting of black and white workers against one another that brought on full-blown disease. That is the essential element that has been ejected from historical discourse by the turn away from a materialist understanding of racism.

White elites did not undertake the industrial transformation of the New South in order to provide a pedestal from which white workers could lord it over poor blacks, as so much contemporary historiography takes for granted. Rather, as Nell Painter has pointed out and as I hope to have demonstrated in this study, the "fundamental point of racism" was "the economic and political domination of the poorest part of the southern working class."[6] That working class was biracial. In various locales throughout the New South (and in the post–World War I period throughout the United States) substantial groups of white workers faced a stark choice. Faced with deliberate attempts to undermine their living and working conditions through the deployment of low-paid, relatively defenseless African Americans, white workers were compelled to choose between linking arms with blacks or attempting to exclude them. Often they attempted the latter. But where exclusion was not a viable option and where other factors (including the presence of white antiracists) favored its development they embraced interracial unionism.

Irrefutably, the experience of linking arms with black strikers in common defense forced white workers to reexamine their relationship with white elites and to reassess their place in the Solid South, the meaning of white supremacy, and the causes of inequality. Whether they were able to move beyond "stomach equality" and toward a broader vision of working-class interracialism was not predetermined beforehand but contingent upon a number of factors. The legacy of white racism proved a powerful impediment to such a development. The traditional separation of politics and economics, so deeply ingrained in the traditions of the American labor movement, did not augur well for a forthright challenge to white supremacy. In many ways the conclusion reached by the best working-class militants and carried into the next round of mass struggles in the 1930s—that the particular, racial oppression of black workers would have to be faced frontally and not sidestepped—was the historical jewel snatched from the rubble of defeated strikes like those that had convulsed the Alabama coalfields between 1916 and 1921.

The essential point, however, is not that opportunities were missed but that here alone—in the day-to-day struggles of working-class southerners—lay the possibilities for a fundamental challenge to racism. The most forward-thinking white southern elites hoped to reduce racial friction while leaving the structures of exploitation and racial oppression intact. Black accomodationists believed that by delivering up ordinary blacks for exploitation in the mines and mills they would win space for themselves and their version of "race progress" within the confines of the Jim Crow South. Warts and all, an interracial UMW aimed at something quite different, in large part because neither black nor white miners had a material stake in perpetuating the arrangement.

Notes

Introduction

1. "It was not . . . race and culture calling out of the South in 1876," Du Bois writes, "it was property and privilege, shrieking to its kind [in the North], and privilege and property heard and recognized the voice of its own" (*Black Reconstruction in America,* 630).

2. Ibid., 700.

3. For recent scholarship on race and the labor movement, see Arnesen, *Waterfront Workers of New Orleans;* Corbin, *Life, Work, and Rebellion in the Coal Fields;* Halpern, *Down on the Killing Floor;* Honey, *Southern Labor and Black Civil Rights* and *Black Workers Remember;* Horowitz, *"Negro and White, Unite and Fight!";* Jones, *American Work;* Letwin, *Challenge of Interracial Unionism;* Lewis, *Black Coal Miners in America;* Minchin, *Hiring the Black Worker;* Stein, *Running Steel, Running America;* and Trotter, *Coal, Class, and Color.* The contours of ongoing, related debates on black workers, race and organized labor, and working-class interracialism can be traced in the following essays, listed in order of their publication: Green and Worthman, "Black Workers in the New South"; Roediger, "'Labor in White Skin'"; Arnesen, "Following the Color Line of Labor"; Halpern, "Organized Labor, Black Workers, and the Twentieth-Century South"; Trotter, "African-American Workers"; Hill, "Problem of Race in American Labor History"; Nelson, "Class, Race, and Democracy in the CIO"; Roediger, "What If Labor Were Not White and Male?"; Lynd, "History, Race, and the Steel Industry"; Norwood, "Bogalusa Burning"; Arnesen, "Up from Exclusion"; Feldman, "Research Needs and Opportunities."

4. Du Bois, *Black Reconstruction in America,* 626–67; Kester, *Revolt among the Share-croppers,* 20.

5. For an illustration of how far contemporary debate has shifted away from the materialist approach, consider David Roediger's rather cursory objection in *The Wages of Whiteness* to Oliver Cromwell Cox's insistence that "economic relations form the basis of modern race relations" (7). Roediger apparently regards the weakness of Cox's argument to be self-evident and offers no substantive explanation for rejecting his prescription that "Blacks and whites should look to class-based revolution as the solution to racism" (7), a conclusion that would have seemed fairly uncontroversial to the generation of activists that I've referred to.

6. For a provocative critique of the "old" labor history—including an engaging discussion of the symmetry between the conservative interpretation developed by John R. Commons and his followers and the left-wing version advanced by the pre-1960s Left, see Merrill, "Interview with Herbert Gutman," 342.

7. Brody, "Old Labor History and the New," 117, 122, 126.

8. Limerick, "Has 'Minority History' Transformed the Historical Discourse?" 1.

9. Fox-Genovese and Genovese, *Fruits of Merchant Capital,* 200, 199, 208, 212; Saville, "Radical Left Expects the Past to Do Its Duty," 273; Callinicos, "Marxism and the Crisis in Social History," 32, 37, 36.

10. Limerick, "Has 'Minority History' Transformed the Discourse?" 1. Limerick argues forcefully that "the formula 'active agents, not passive victims' has unintentionally proven a good way to draw attention away from the agency, name, and responsibility of those who accumulated coercive power and used it to injure culturally enriched but economically ripped-off groups" (34). Though he rejects class relations as a viable alternative framework for interpreting the African-American past and is curiously uncritical of the primacy of "culture" in understanding "whiteness," Clarence E. Walker makes a similar point with regard to developments in his field. He objects that "new studies of black life in slavery and afterward gave culture a power and determinative force that [the historical record provides] no *prima facie* reason for allowing" (*Deromanticizing Black History,* xii). The "romanticism that currently characterizes the writing of black history," he continues, "is really part of a problem inherent in the new social history. In their efforts to recreate a world we have lost, social historians have shed light on groups that were previously thought to be without history. But in accomplishing this goal they appear reluctant to apply the same standards of critical evaluation to 'the people' that they apply to elites" (xviii).

11. Gutman, "Negro and the United Mine Workers," 117, 100, 115.

12. Hill, "Myth-Making as Labor History," 132–99 (quotes on 136, 132–33).

13. Gutman, "Negro and the United Mine Workers," 110, 59, 65; Hill, "Myth-Making as Labor History," 135.

14. On his faith in the ameliorative potential of southern capital, see Hill, "Recent Effects of Racial Conflict on Southern Industrial Development." Sections of southern industry—particularly new industries less dependent on what Glen Eskew calls the "racial wage"—did, in the final days of Jim Crow, seek a compromise that would damp down the protest unleashed by the civil rights movement, but in the earlier period, when Hill was writing, they provided the backbone for the segregationist resistance. Bartley reports that "the Southern States Industrial Council, which claimed five thousand member businesses, along with the Arkansas Free Enterprise Association, the Associated Industries of Florida, and a few other state business associations, played an active role in the [states' rights] campaign. [These] groups served . . . smaller enterprises, many of them in traditional southern industries, for which cheap labor had a telling effect on the balance sheet" (*New South,* 83). On the split in Birmingham industry later on, see Eskew, *But for Birmingham,* 153–92.

15. Conspicuously, for a scholar concerned with dissecting race relations in steel, McKiven devotes just three paragraphs in *Iron and Steel* to white steelworkers' belated effort in 1918 to assist the organization of unskilled blacks. On that occasion Birmingham's steel

interests unleashed armed vigilantes on black unionists, breaking up meetings and tar-and-feathering a black Mine Mill organizer before driving him out of town. McKiven devotes only three pages to TCI's nationally renowned company welfare policy and understates the importance of antiunionism in the company's relations with black workers.

16. Hill, "Problem of Race in American Labor History," 192; McKiven, *Iron and Steel,* 2; Draper, *Conflict of Interests,* 15. In comments at a panel on "Race-ing Southern Industry: Railwaymen, Tobacco Workers, and Racial Boundaries in the American South" at the 1997 Southern Labor Studies Conference in Williamsburg, Virginia, Draper spoke of the need to "bring the employers back in" to the debate over race and the labor movement, but he implied that they had played a progressive rather than an obstructionist role in helping black workers secure their civil rights (author's notes, Sept. 26, 1997).

17. Arnesen writes, perceptively, that the "current rage to demonstrate the social construction of race and white workers' agency in creating their own racism has let capital largely off the hook, with workers dividing themselves and capital merely walking away with the proverbial shop" ("Up from Exclusion," 156). The most influential "whiteness" studies generated in recent years include Roediger, *Wages of Whiteness* and *Towards the Abolition of Whiteness;* and Ignatiev, *How the Irish Became White.*

18. Lynd, "History, Race, and the Steel Industry," 14–15.

19. George Gordon Crawford, a Tennessee Coal, Iron, and Railroad executive, cited in Leighton, *Five Cities,* 129.

20. Strong parallels link the much more heavily studied mill village paternalism that suffused the southern textile industry and that which developed in the mines of the Birmingham district, but the racial composition of mining—particularly in the Alabama fields—made for a very different dynamic, one in which the employers' ascendancy depended in part on their ability to manipulate workforce racial antagonisms. The promotion of the mills as the salvation of poor whites and the exclusion of blacks from production work in textiles did not remove race as a factor in the industry, but it did rule out the possibility of even limited interracial collaboration. For an overview of the literature on textile paternalism, see Carlton, "Paternalism and Southern Textile Labor." He points to the "increasing realization that industrial paternalism had deep American, not just southern, roots and that, far from being associated with 'retrograde' social institutions such as the slave plantation, it was intertwined with the origins of modern capitalism itself" (19). While accepting that paternalism was not exclusive to the South and that it could develop exclusive of any ties with slavery, I argue throughout this study that the context in which industrialization occurred in Birmingham (what Mary Lethert Wingerd has termed the "specific contingencies of personality and place" [*Rethinking Paternalism,* 874])—including the cultural residue of slavery, the central role of black labor, and the direct involvement of planters in early management—were critical factors giving shape to the peculiar form of paternalism embraced by district operators. See my article "Policing the 'Negro Eden'" for an elaboration of this argument. A closer match to the variant of paternalism found in coal seems to have developed in sections of the southern lumber industry. See Fickle, "Management Looks at the 'Labor Problem'"; and Sitton and Conrad, *Nameless Towns,* 79–126.

21. Brier, "In Defense of Gutman," 389. In relation to the UMW, Hill writes that "especially after the 1898 strikes in Illinois, to insist upon excluding blacks from the [UMW] was

impossible" ("Myth-Making as Labor History," 160). Concerning the CIO, he stresses that "blacks were admitted into CIO unions because it was in the self-interest of whites to do so." He is correct, of course, that interracialism was far more often forced upon white workers than deliberately pursued and that mixed union locals fell short of practicing thoroughgoing racial equality, a fact that seems unremarkable in the context of the early twentieth-century United States, when few civic organizations of any kind practiced even token integration. Even if in some sense involuntary, the shift to mixed or integrated locals was itself significant, opening up the possibilities for racial egalitarianism ruled out in labor organizations that maintained the color bar or held to a policy of exclusion. Whether "fundamental shifts in institutional arrangements" occurred within unions was contingent on a number of factors, including whether or not a core of white unionists—whether out of political motivations or otherwise—were willing to fight for such an outcome. His insistence that "no less for communist-controlled unions . . . than for the rest of organized labor . . . the prevalence of white racism overwhelmed the few scattered examples of interracial unionism" is unsustainable ("Problem of Race in American Labor History," 199, 202).

22. Saville, "Radical Left Expects the Past to Do Its Duty," 273. The classic summary of the materialist interpretation of history appears in Marx, "Eighteenth Brumaire of Louis Bonaparte," 11:103.

23. Stein, *World of Marcus Garvey,* 56. See also Arnesen, "Following the Color Line of Labor," where he argues that "in an era of massive labor unrest, southern black working-class activity was central to the era's African-American and labor history. Notwithstanding the advice of conservative black leaders who propagated Booker T. Washington's pro-industrialist philosophy, black workers were hardly strangers to classic forms of class conflict" (71–72). Stein argues that "the issues of jobs, wages, rent, and food prices were forced on the new agitators only after the end of the war when the inevitability of racial progress in industrial society, in the North as well as in the South and in Africa, was called into question" (54) and in a situation where "the racial elite was caught in the middle of escalating black demands and southern efforts to restore the prewar social order" (57). Her argument that the NAACP was able to transform itself during this period in large part because of its ability to move "to the left, towards working-class methods of protest" (58), has been confirmed in important recent studies that link increased proletarianization and rising black militancy. See, for example Reich, "Soldiers of Democracy"; Obadele-Starks, "Black Labor, the Black Middle Class, and Organized Protest along the Upper Texas Gulf Coast"; and the essays by various contributors in Trotter (ed.), *Great Migration in Historical Perspective.*

24. On strike activity during the war and in the immediate postwar period, see Foner, *Labor and World War I* and *Postwar Struggles,* and Montgomery, *Fall of the House of Labor,* 330–410. On anti-immigrant and antiradical hysteria, see Preston, *Aliens and Dissenters,* 88–238.

25. See Halpern, *Down on the Killing Floor,* especially his excellent account of the 1919 Chicago Race Riot (56–72).

26. Montgomery, *Fall of the House of Labor,* 370.

27. See <H-NET@msu.edu> (Aug. 1998): hreview@h-net.msu.edu for my review of Letwin's study, which expands on a number of points made later.

28. Halpern, "Organized Labor, Black Workers, and the Twentieth-Century South," 75.

29. Letwin rightly insists that "alternating periods of ascent and defeat turned not so much on a waxing and waning of interracial cooperation as on an evolving constellation of external circumstances," among them the "superior resources of the operators" (*Challenge of Interracial Unionism,* 7). Interracial cooperation was, in other words, a necessary precondition for mounting a challenge to the operators, but it was by no means a guarantee of success.

30. Ibid., 54, 89, 140.

31. Bayard Rustin, cited in Gatewood, "Aristocrats of Color, South and North," 3.

32. In his discussion of the relationship between black miners and Birmingham's conservative black middle class, Letwin cites a speech from the black UMW vice president B. L. Greer to support the argument that "the philosophies (or at least rhetorical strategies) of the black middle class and black unionists in the New South were not always so incompatible as is often supposed" (*Challenge of Interracial Unionism,* 145). Yet it is important to distinguish between rhetoric and substance here. Greer's admittedly curious assertion that blacks had "been given suffrage, it is said, too soon" was likely intended to express very different concerns than those voiced by Washington and his followers. Washington's embarrassment regarding the franchise derived from his frustration that so many blacks had deserted the "better class of whites" to cast their votes along with white plebeians in the Populist and anti-Redeemer movements. It is almost certain that Greer, a veteran unionist, viewed the problem from a very different perspective: more likely he shared the sentiment, widespread among ex-Populists, that too many blacks had been duped into voting for Bourbon reactionaries.

Prologue

1. *Birmingham Age-Herald,* Aug. 8, 1908.

2. On black support for Evans in the 1896 election, see Harris, *Political Power in Birmingham,* 68.

3. Meaning those companies that produced coal for use in their own blast furnaces as opposed to selling it on the commercial market. In Alabama these captive mine operators dominated the local coal industry.

4. This system benefited operators in a number of ways. It left responsibility for meeting tonnage rates in the hands of subcontractors, many of whom were not miners themselves, and it relieved employers of any liability for death or injuries among mine "laborers." The contracting system also reinforced the racial division of labor in the mines. Characterizing the practice as "distinctly and absolutely a product of non-unionism," the prominent Birmingham journalist Ethel Armes observed in an exposé in the *Birmingham Labor Advocate* (Apr. 18, 1913) that subcontractors typically employed "anywhere from four to five to as many as 25 laborers, for the lowest wage they will work for." Furthermore, "he gets a handsome profit clear every month, off of the labor of his negroes or his poor whites—for only the poorer class . . . in this district 'stays in' with a sub-contractor." Elsewhere Armes states that "many a farmer . . . brought his hired hands along—negroes who worked for what he elected to pay."

5. *Birmingham Age-Herald,* July 12, 1908.

6. See Straw, "'This Is Not a Strike, It Is Simply a Revolution,'" 70–73.

7. *Birmingham Age-Herald,* July 30, 1908, Aug. 4, 1908.

8. *New York Times,* Aug. 11, 1908; *Birmingham Age-Herald,* July 23, 1908; "Interview: John Sokira," 3, Oral History Collection, Special Collections, Samford University, Birmingham, Ala.

9. *New York Times,* July 18, 1908; *Birmingham Age-Herald,* July 18, 1908, July 30, 1908, Aug. 20, 1908, Aug. 10, 1908.

10. *Birmingham Age-Herald,* Aug. 10, 1908; Straw, "'This Is Not a Strike, It Is Simply a Revolution,'" 94; *Birmingham Age-Herald,* Aug. 25, 1908.

11. *Birmingham Age-Herald,* Aug. 24, 1908, Aug. 22, 1908, Aug. 25, 1908; Hornady, *Book of Birmingham,* 55; *Birmingham Age-Herald,* Aug. 27, 1908.

12. Lewis, *Black Coal Miners in America,* 49; *Birmingham Age-Herald,* July 21, 1908, July 18, 1908, Aug. 20, 1908.

13. *Birmingham Age-Herald,* Aug. 14, 1908, Aug. 1, 1908, Aug. 13, 1908.

14. Ibid., Aug. 25, 1908, July 30, 1908.

15. Ibid., Aug. 3, 1908.

16. Ibid., Aug. 5, 1908, Aug. 18, 1908; Straw, "'This Is Not a Strike, It Is Simply a Revolution,'" 106.

17. *Birmingham Age-Herald,* Sept. 25, 1908, Aug. 29, 1908; Straw, "'This Is Not a Strike, It Is Simply a Revolution,'" 105.

18. *Birmingham Labor Advocate,* Sept. 25, 1908.

Chapter 1: The Operators' Dilemma

1. *Birmingham Labor Advocate,* Sept. 4, 1908; Straw, "'This Is Not a Strike, It Is Simply a Revolution,'" 250; "Interview: Samuel Kelley and James Simmons," 5–6, Oral History Collection, Special Collections, Samford University, Birmingham, Ala.; John P. White quoted in *Birmingham Labor Advocate,* Sept. 25, 1908.

2. *Birmingham Ledger,* Sept. 9, 1908; *Birmingham Age-Herald,* Sept. 3, 1908; *Birmingham Labor Advocate,* Oct. 2, 1908, Sept. 28, 1908.

3. The ACOA president George B. McCormack of Pratt Consolidated Coal Company recalled in his 1912 annual report that the "association was formed in 1908 for the purpose of defending ourselves against the unreasonable demands of the Coal Miners' Union." See "1912 President's Annual Report," Alabama Coal Operators' Association/ Alabama Mining Institute Records (hereafter ACOA/AMI Records), Department of Archives and Manuscripts, Birmingham Public Library, Birmingham, Ala.

4. *Birmingham Labor Advocate,* Oct. 2, 1908; Fitch, "Birmingham District," 1538, and "Executive Board Minutes," Sept. 24, 1908, Feb. 11, 1909, June 3, 1909, ACOA/AMI Records.

5. *United Mine Workers' Journal,* Sept. 24, 1908. On blacklisting after the 1908 strike, see "Interview: Kelley and Simmons," 6; "Executive Board Minutes," Sept. 24, 1908, ACOA/ AMI Records; "Dan J. DeJarnette to Van Bittner," Dec. 13, 1920, Van Amburg Bittner Papers (hereafter Bittner Papers), West Virginia and Regional History Collection, University of West Virginia, Morgantown; *Birmingham Labor Advocate,* Sept. 4, 1908.

6. *Wall Street Journal,* May 16, 1905.

7. "1910 President's Annual Report," June 7, 1910, and "1911 President's Annual Report," June 6, 1911, both in ACOA/AMI Records; U.S. Immigration Commission, *Immigrants in Industries,* 363; *Birmingham Labor Advocate,* July 22, 1910; Fitch, "Birmingham District," 1538; Moran et al., "Report of the Committee on Conditions in the Alabama Coal Mines,

1913," cited in Taft, *Organizing Dixie*, 46; see also "Report on District 20 14th Annual Convention," June 12, 1911, in Philip Taft Research Notes on Alabama Labor History, 1902–77 (hereafter Taft Research Notes), Department of Archives and Manuscripts, Birmingham Public Library, Birmingham, Ala.

8. Edwin C. Eckel, reprinted from the *Manufacturers' Record*, in *Birmingham Age-Herald*, Sept. 2, 1907.

9. *United Mine Workers' Journal*, Sept. 24, 1908. A high proportion of those who left the district after the 1908 defeat were skilled miners, which meant that, given the racial division of labor in Alabama mines, they were also disproportionately white. Durrett's letter appears in *Birmingham Labor Advocate*, Oct. 30, 1908.

10. Dodge was apparently a seasoned labor agent by this time. He had been shot in the leg and hand while escorting a trainload of strikebreakers at Blocton in August 1908. See *Birmingham Age-Herald*, Aug. 10, 1908.

11. On the importance of Jim Crow legislation for maintaining a cheap labor supply, see Green and Worthman, "Black Workers in the New South," 48. The organization of the ACOA's Labor Bureau is detailed in "Executive Board Minutes," Nov. 6, 1909, ACOA/AMI Records.

12. Text of the letter to railroad agents in "Executive Board Minutes," Apr. 9, 1912, ACOA/AMI Records.

13. *Birmingham Labor Advocate*, Apr. 21, 1911.

14. Ibid., July 22, 1910, Dec. 11, 1908.

15. Lewis, *Sloss Furnaces and the Rise of the Birmingham District*, xiv; Leighton, *Five Cities*, 116.

16. Lothian Bell quoted in Armes, *Story of Coal and Iron in Alabama*, 301; Carnegie quoted in Wright, *Old South, New South*, 165; Armes, *Story of Coal and Iron in Alabama*, 467; Leighton, *Five Cities*, 100; Herbert N. Casson, "Alabama District to Make Millionaires by the Score," *Munsey Magazine*, Feb. 1907, reprinted in *Birmingham Age-Herald*, Jan. 28, 1907.

17. *Birmingham Age-Herald*, Nov. 6, 1907; *Birmingham News*, Nov. 6, 1907, in Armes, *Story of Coal and Iron in Alabama*, 521; *Birmingham Labor Advocate*, Mar. 31, 1906.

18. Sanford letter cited in Lewis, *Sloss Furnaces and the Rise of the Birmingham District*, 294. On northern domination of southern steel and the effects of "Pittsburgh Plus," see also Wright, *Old South, New South*, 169–70; Leighton, *Five Cities*, 122–30; Harris, *Political Power in Birmingham*, 18–20; Woodward, *Origins of the New South*, 302, 315.

19. At least one prominent economic historian has argued that in the long run U.S. Steel's pricing policy may have had a beneficial effect on the late-developing southern steel industry by providing protection from national and international competition during its crucial early years. See Wright, *Old South, New South*, 170.

20. On metallurgical problems, see Fuller, "From Iron to Steel," 138, and Cushing, "Birmingham Coal Puzzle," 1–2; Wiebel, *Biography of a Business*, 48; Lichtenstein, *Twice the Work of Free Labor*, 79–80; "Testimony of J. D. Kirkpatrick before the Kilby Commission," ACOA/AMI Records; *Birmingham Age-Herald*, Nov. 18, 1907. On the weakness of home markets and the effects of high shipping costs, see Armes, *Story of Coal and Iron in Alabama*, 303.

21. Grady, "South and Her Problems," 46–47.

22. Woodward, *Origins of the New South,* 307. For a classic example of the New South appeal to northern capital, see the article from the *Manufacturer's Record* cited in Leighton, *Five Cities,* 129. On the importance of cheap labor to southern industrialization, see Cobb, *Industrialization and Southern Society,* 24–26. David Montgomery writes that "the [South's] industrial growth was founded on extractive and manufacturing enterprises, which could yield a good return . . . only if the wages paid were much lower than those in the North" (*Fall of the House of Labor,* 84).

23. Fitch, "Birmingham District," 1529, 1531; U.S. Department of Labor, *History of Wages in the United States from Colonial Times to 1928,* D-3:228–9, G-1:260.

24. On the importance of the open shop to the Alabama coal industry, see "1918 President's Annual Report," ACOA/AMI Records; Richards, "Racism in the Southern Coal Industry," 8–12; Lewis, *Black Coal Miners in America,* 48; Letwin, "Race, Class, and Industrialization," 5, 80; Wilson, "Structural Imperatives behind Racial Change in Birmingham, Alabama," 179–80; Lichtenstein, *Twice the Work of Free Labor,* 79–80; Dong, "Social and Working Lives of Alabama Coal Miners and Their Families," 14, 41–42, 93.

25. Cushing, "Birmingham Coal Puzzle," 2–5.

26. Statistics on mechanization and labor costs in the coal industry from Richards, "Racism in the Southern Coal Industry," 8–12, esp. tables 3 and 4. On the retarding effects of low wages on mechanization in the Birmingham district, see Menzer and Williams, "Images of Work," 11. See also Lewis, "Emergence of Birmingham as a Case Study of Continuity between the Antebellum Planter Class and Industrialization in the 'New South,'" where he demonstrates the drag exerted on technological innovation in the South by the availability of cheap labor with an illustration from the pig-iron industry: "While [Sloss-Sheffield] did not utilize convicts at its blast furnaces, as it did in its mining camps, it clung to time-honored manual techniques by persisting in employing low-paid black workers for such arduous tasks as casting pig iron. Paradoxically, a highly sophisticated automatic pig-casting machine was conceived at Birmingham in the early 1890s by Edward A. Uehling, a northern engineer . . . appointed to supervise [the] blast furnaces [but] because the device was expensive and ill-suited to labor-intensive southern furnace practice, it had no chance of being adopted in Alabama at the time. . . . Uehling sold his patent rights to Andrew Carnegie and the machine became part of the 'Duquesne Revolution' that took place in the Pittsburgh District. Not until the 1920s and early 1930s did Sloss-Sheffield gradually adopt mechanical casting" (76).

27. U.S. Department of Labor, *History of Wages in the United States from Colonial Times to 1928,* J-2:329, J-6:332, J-4:330, J-11:335. Crawford cited in Leighton, *Five Cities,* 129. Alabama operators' ability to undercut unionized mines in the Central Competitive Field made it more urgent that the UMW organize the district: "rising volume of coal from the South," Montgomery writes, "made CCF operators increasingly restive. In 1904 they demanded and won a wage reduction as the price of continuing [their] agreement [with the UMW]" (*Fall of the House of Labor,* 343).

28. "Output of Alabama Coal Operators, 1911," in "Annual Report of the Mine Inspector for the State of Alabama for 1912," 532, ACOA/AMI Records.

29. On differences between the work process and skill requirements in steel, pig iron, and coal, see Lewis, *Sloss Furnaces and the Rise of the Birmingham District,* 314–15, and Fitch, "Birmingham District," 1527–29.

30. Fies, "Industrial Alabama and the Negro: Speech before the Alabama Mining Institute," Oct. 31, 1922, ACOA/AMI Records.

31. Lewis, *Black Coal Miners in America,* 10–12; Harris, *Political Power in Birmingham,* 108; "J. D. Kirkpatrick Testimony Before the Kilby Commission," ACOA/AMI Records; Aldrich's testimony in U.S. Senate, *Relations between Labor and Capital,* 384–85, cited in Letwin, "Race, Class, and Industrialization," 55.

32. Fitch, "Birmingham District," 1527.

33. *Birmingham Age-Herald,* Aug. 16, 1903; James W. Sloss cited in Letwin, "Race, Class, and Industrialization," 55; *Birmingham Age-Herald,* Jan. 28, 1907. See also Gibbs, "Negro as an Economic Factor in Alabama," 48, and Bigelow, "Birmingham," 127.

34. Fitch, "Birmingham District," 1527; *Birmingham Age-Herald,* Jan. 28, 1907; Harrison, "A Cash Nexus for Crime," 1546; *Birmingham Age-Herald,* Jan. 28, 1907.

35. Lewis, in *Sloss Furnaces and the Rise of the Birmingham District,* reports that "in the early years of the [twentieth] century, TCI recruited 5,000 steelworkers and coal miners from Italy alone" (315).

36. *Birmingham Age-Herald,* Jan. 17, 1907, Jan. 28, 1907, June 23, 1907, Jan. 21, 1907.

37. Ibid., Aug. 21, 1907.

38. Ibid., Aug. 21, 1907, Aug. 11, 1907.

39. "Executive Board Minutes," Sept. 7, 1910, ACOA/AMI Records. In an important study of southern lumber operators' attitudes toward black labor, James E. Fickle finds that in the years straddling World War I the main employers' organization acknowledged that "the South was not a paradise for blacks." "Even though [they] held the traditional southern white view of the Negro," he writes, an industry-wide survey "revealed the conviction within the industry that Negroes were valuable to the South, necessary to the operation of the southern pine industry, and suffering from many undesirable and even intolerable conditions." See Fickle, "Management Looks at the 'Labor Problem,'" 73.

40. "Report of the September 1911 Grand Jury of the Criminal Court of Jefferson County," Classified Subject Files, 1914–41, #50–112, Records of the Department of Justice, RG 60, National Archives, Washington, D.C.; "Petition to the Excellent Governor O'Neal," July 7, 1914, Governor Emmett O'Neal Administrative Files, Alabama Department of Archives and History (hereafter ADAH), Montgomery.

41. Harris reports that "as a remedy against the supposed Negro shiftlessness, employers frequently demanded stricter vagrancy law enforcement. . . . Periodically between 1890 and 1920 city police mobilized to 'wage unrelenting war against the vagrants,' under such slogans as 'no idleness will be tolerated. . . . It is either a case of go to work or go to jail.'" Significantly, the "anti-vagrant drives against sheer idleness concentrated almost entirely on Negroes." Harris, *Political Power in Birmingham,* 199–200.

42. Leighton reports that the office of Jefferson County sheriff by this period had become "the most lucrative office in the state with a fee income of between fifty and eighty thousand dollars" annually (*Five Cities,* 118). For a trenchant analysis of employer opposition to the fee system, see Harris, *Political Power in Birmingham,* 207–15. Quotes from Harris, *Political Power in Birmingham,* 210; Walker Percy of TCI in *Birmingham News,* Jan. 6, 1911, Jan. 14, 1911; Harris, *Political Power in Birmingham,* 212. One investigation charged that Sheriff Higdon of Jefferson County had "for a certain sum of money . . . entered into an agreement or compact with the gamblers that he would allow them to openly violate

the laws and furnish protection to them." See *Birmingham Labor Advocate,* Dec. 18, 1908. Union miners alleged that companies profited from paying their employees' fines, charging black miners up to 25 percent interest on fees and court costs, to be deducted from their pay. See *Birmingham Labor Advocate,* Jan. 13, 1911.

43. *Birmingham Age-Herald,* Sept. 26, 1907, Aug. 25, 1907, Aug. 24, 1907. For antiprohibition advertisements, see the *Birmingham News,* Sept. 18, 1907, *Birmingham Age-Herald,* Aug. 25, 1907, and Oct. 21–28, 1907. On working-class opposition to prohibition, see the *Birmingham Labor Advocate,* 1906–7; on prohibition in 1907 generally, see Harris, *Political Power in Birmingham,* 81, 194.

44. Harris, *Political Power in Birmingham,* 35; Fitch, "Birmingham District," 1527.

45. *Birmingham Age-Herald,* Aug. 18, 1907; Xaridimos to Stefane, June 12, 1909, "Letter from Greek Writer, West Blockton [*sic*], Alabama: Complaint of Apparent Peonage upon Part of TCI," Peonage Files, Records of the Department of Justice, RG 60.

46. Cobb, *Most Southern Place on Earth,* 110. On segregation of immigrants in the coal camps, see "Mary Worked in the Mines in Belgium," in Federal Writers' Project, *Alabama,* 470–71, and "Interview: John Sokira," 8; Berthoff, "Southern Attitudes toward Immigration," 349. Neither Birmingham nor the South generally held a monopoly on anti-immigrant prejudice, of course. David Montgomery points out that "American culture at the turn of the century categorized all the strangers who filled the laborers' ranks as 'lesser breeds,' except those who were native white Americans, and they were 'bums'" (*Fall of the House of Labor,* 81).

47. *Birmingham Age-Herald,* Jan. 15, 1907, Oct. 21, 1907.

48. Fitch, "Birmingham District," 1527; *Birmingham Age-Herald,* Dec. 6, 1907; S. T. Stephens to Department of Justice, June 14, 1909, Classified Subject Files, 1914–41, #50–95–0, Records of the Department of Justice, RG 60.

49. Wright, *Old South, New South,* 76–77; Berthoff, "Southern Attitudes toward Immigration," 357; *Birmingham Labor Advocate,* Nov. 12, 1909.

50. Lewis reports that as of 1913, "immigrants [at TCI] reported for work about twenty days per month, blacks not quite seventeen. For operatives as a whole the average was just over eighteen, indicating that TCI's native whites were not well disciplined either" (*Sloss Furnaces and the Rise of the Birmingham District,* 316). See also *Birmingham Age-Herald,* Jan. 28, 1920. Fitch explained that "industry is young in all the South" and "there has not been time for the development of a body of skilled labor in Alabama . . . it is also true that the southern whites who come to work in the mills and mines are generally of the poorer sort" ("Birmingham District," 1528).

51. Correspondence from Alabama miners throughout the first two decades of the twentieth century is filled with complaints about hardship caused by the short workweek. Nov. 15, 1907: "The mines at Fossil closed down last evening, and that means that about 300 men are thrown out of work. The spirit of unrest is very great in the city just at present, although every one seems to think that conditions will be better in the future." July 18, 1908: "The company has not given the poor men enough to hardly maintain his family for they sell things so high in the commissary and for two or three months the company has just let our poor men have two or three days work a week" ("Mrs. R. H. Cot to Governor Comer," Governor Braxton Bragg Comer Administrative Files, ADAH). Aug. 4, 1911: "Mines running four and five days a week." Aug. 8, 1913: "Work is still slack; four days a week." Oct. 28, 1914: "These mines have worked five and one-half days this month. . . .

All other mines in this state with but few exceptions have done no more, but the convicts . . . are working every day and producing nearly 100,000 tons of coal per month." Nov. 13, 1914: "The mine at Sayre has been closed for months." Dec. 18, 1914: "The miners, as usual, worked two days last week." Dec. 25, 1914: "We have worked seven and one-half days so far this month, so you can guess how Christmas will be here." All quotes from the *Birmingham Labor Advocate* unless otherwise noted.

52. "Interview: Luther V. Smith," 8, and "Interview: Elmer Burton," 1, Oral History Collection, Special Collections, Samford University, Birmingham, Ala.; U.S. Senate, *Relations between Labor and Capital,* cited in Wright, *Old South, New South,* 98.

53. One additional caveat is important in considering the question of mobility. At least one influential study has shown that, contrary to the employers' complaints, blacks in late nineteenth-century Birmingham were actually the least mobile of all industrial employees and that among white workers unskilled laborers were less mobile than skilled workers. See Worthman, "Working Class Mobility in Birmingham, Alabama," 185. Genovese, *Roll, Jordan, Roll,* 298–89. See also Edward P. Thompson's fascinating essay, "Time, Work, and Industrial Capitalism," in *Customs in Common,* 352–403. Thompson finds that in the mid-nineteenth century "it was commonly observed that the English industrial worker was marked off from his fellow Irish worker, not by a greater capacity for hard work, but by his regularity, his methodical paying-out of energy, and perhaps also by a repression . . . of the capacity to relax in the old, uninhibited ways." He cites an early twentieth-century assessment of the problems with Mexican mine labor that could easily have been uttered by a Birmingham operator in relation to blacks during the same period: "'His lack of initiative, inability to save, absences while celebrating too many holidays, willingness to work only three or four days a week if that paid for necessities, insatiable desire for alcohol—all were pointed out as proof of natural inferiority'" (396–97).

54. On the ACOA's abandonment of the Central Labor Bureau, see "1910 President's Annual Report," June 7, 1910, ACOA/AMI Records. On the shooting of labor agents at Corona, see *Birmingham Age-Herald,* Aug. 20, 1907.

55. "Handling the Negro Miner in the South," 875.

56. Llewelyn Johns, quoted in Letwin, "Race, Class, and Industrialization," 57. On TCI's early attempts at raising standards among black workers, see Lewis, *Sloss Furnaces and the Rise of the Birmingham District.* Lewis argues convincingly that TCI's "first steps toward welfare capitalism" and its aggressive lead in attempting the immigration campaign corresponded with the company's "shift into basic steelmaking" (315).

57. Topping quoted in Davis, *Labor and Steel,* 147.

58. Wiebel, *Biography of a Business,* 44–45, 41.

59. Tarbell, *Life of Elbert H. Gary,* 310–11.

60. Carmer, *Stars Fell on Alabama,* 80, 81.

Chapter 2: The Limits of Reform from Above

1. Lewis, *Sloss Furnaces and the Rise of the Birmingham District,* 316.

2. *Wall Street Journal,* May 16, 1905.

3. "1914 President's Annual Report," "1912 President's Annual Report," ACOA/AMI Records.

4. Bigelow, "Birmingham," 3–4. On U.S. Steel's protracted battle with organized labor, see Garraty, "United States Steel Corporation versus Labor," 3–38.

5. Richards, "Racism in the Southern Coal Industry," 33, 32.

6. Gary quoted in Knowles, "Relation of the Operator to Sanitary and Social Improvements," July 11, 1912, *Publications of the ACOA*, 5, ACOA/AMI Records, and in Lewis, *Sloss Furnaces and the Rise of the Birmingham District*, 317.

7. "Testimony of John A. Topping," U.S. Senate, Committee on Investigation of the United States Steel Corporation, *Hearings on the United States Steel Corporation* (hereafter *U.S. Steel Hearings*), vol. 2, 1269.

8. "Testimony of Don H. Bacon," *U.S. Steel Hearings*, vol. 7, 5333, 5346–47; "Testimony of John A. Topping," *U.S. Steel Hearings*, vol. 2, 1228, 1269.

9. Davis, *Labor and Steel*, 146.

10. The TCI manager George Gordon Crawford brought new enthusiasm to the company's welfare plans after his appointment in 1907. Lewis reports that Crawford "hired professional medical and social workers to administer a growing network of clinics, schools, and recreation facilities. In 1908, he appointed the prominent black banker and educational reformer [and Booker T. Washington protégé] W. R. Pettiford to develop an instructional program in housekeeping, sewing, and other domestic skills at TCI's mining camps." For a full discussion of Crawford's contribution at TCI, see Lewis, *Sloss Furnaces and the Rise of the Birmingham District*, 318–21, and Rikard, "George Gordon Crawford," 163–81.

11. "1912 Secretary-Treasurer's Annual Report," July 12, 1913, "Executive Board Minutes," June 6, 1911, ACOA/AMI Records.

12. Tarbell, *Life of Elbert H. Gary*, 310–11.

13. Bowron, "Convicts in Coal Mines in Alabama," 77, 78.

14. Lichtenstein, *Twice the Work of Free Labor*, 95.

15. For a documentary record of the conflict between TCI and the State of Alabama over the removal of convicts from TCI's mines, including the correspondence between TCI officials and the state, see "Testimony of President Oakley, Alabama State Board of Convicts," *U.S. Steel Hearings*, vol. 4, 3112 (quote from Oakley on 3111; excerpts from Crawford letter on 3112). On low lease rates after 1908, see Harris, *Political Power in Birmingham*, 206, and Ward and Rogers, *Convicts, Coal, and the Banner Mine Tragedy*, 114–15.

16. Crawford letter in *U.S. Steel Hearings*, vol. 4, 3112. Crawford was, in fact, forced to construct new housing by the refusal of free miners to occupy buildings that had formerly housed TCI's convicts.

17. Fitch, "Birmingham District," 1540, 1532. Interestingly, another of Fitch's articles, written for the *American Magazine* less than a year before the Birmingham exposé, had provoked an internal fight within the ranks of U.S. Steel. See Garraty, "United States Steel Corporation versus Labor," 33.

18. The four renegade companies were all among the state's top twenty (of 212) producers in 1911, with a combined annual tonnage of nearly 15 percent of the state's output. See "Output of Alabama Coal Operators, 1911," in the *Annual Report of the Mine Inspector for the State of Alabama for 1912*, 532, ACOA/AMI Records. Moore and Maben quoted in *Birmingham Age-Herald*, Jan. 14, 1912; J. W. McQueen quote on "new-fangled ideas" cited in Lewis, *Sloss Furnaces and the Rise of the Birmingham District*, 322; McQueen resignation and withdrawal of coal operators from ACOA noted in "Executive Board Minutes," Jan. 27, 1912, ACOA/AMI Records.

19. Roden, "Value of Improved Sanitary and Living Conditions in Mining Camps,"

Publications of the ACOA, 34, ACOA/AMI Records; "Testimony of John Fitch," *U.S. Steel Hearings,* vol. 4, 2943. The ACOA president G. B. McCormack corresponded with Fitch prior to publication of the special edition of *Survey.* Apparently Fitch had submitted a draft of his article to the major companies in the district and offered them an opportunity to dispute his findings before the piece went to press. See "Executive Board Minutes," Sept. 28, 1911, ACOA/AMI Records; "Testimony of John Fitch," *U.S. Steel Hearings,* vol. 4, 2942. In the end, Fitch's suggestion that Shelby Harrison be contracted for the work was vetoed by TCI's Frank Crockard, who favored Knowles.

20. Fitch, "Birmingham District," 1532.

21. *Birmingham Labor Advocate,* July 18, 1913, July 25, 1913.

22. Ibid., July 25, 1913, Nov. 13, 1914.

23. Ibid., Aug. 21, 1914, May 21, 1915.

24. Lewis, *Sloss Furnaces and the Rise of the Birmingham District,* 323, 322.

25. Fitch, "Birmingham District," 1540.

26. Lewis, *Sloss Furnaces and the Rise of the Birmingham District,* 329.

27. Fies, "Man with the Light on His Cap," 10, 5; and Fies to Henry T. DeBardeleben, June 24, 1920, DeBardeleben Coal Company Records (hereafter DeBardeleben Records), Department of Archives and Manuscripts, Birmingham Public Library, Birmingham, Ala.; Flynt, *Poor but Proud,* 117–18; "Annual Report of the Vice President," 1918, Stockholders' Meetings, vol. 3, Records of the Alabama Fuel and Iron Co. (hereafter AFICo Records), Hoole Library Special Collections, University of Alabama, Tuscaloosa.

28. "Little Cahaba Mine—Piper" and "Barney Coal Company—Cordoba," Mining Community "A" Schedules with Camp Ratings, 1922–23, Records of the Division of Labor Facts and Living Conditions, Records of the Coal Commission, RG 68, National Archives, Washington, D.C.

29. "1912 Secretary-Treasurer's Annual Report," July 12, 1913, ACOA/AMI Records; "Relation of the Operator to Sanitary and Social Improvements," *Publications of the ACOA,* 3, and "1912 Secretary-Treasurer's Annual Report," ACOA/AMI Records.

30. In a speech before the annual meeting of the ACOA in 1911, the TCI official Frank Crockard acknowledged that "there is not a single company in Alabama which is conforming to [ideal welfare standards], some are trying to do so but we crawl before walking and large bodies move slowly even when walking." See *Birmingham Age-Herald,* July 30, 1911. Quote from Lewis, *Sloss Furnaces and the Rise of the Birmingham District,* 296. Brandes argues that welfare capitalism was "affected by the progressive milieu . . . while managing to remain apart from it" (*American Welfare Capitalism,* 37). On the connections between company welfare in the Birmingham district and welfare capitalism nationally, see Lewis, *Sloss Furnaces and the Rise of the Birmingham District,* 321.

31. "1912 Secretary-Treasurer's Annual Report," July 12, 1913, ACOA/AMI Records; Nolan, "Welfare Work of the Tennessee Coal, Iron, and Railroad Company," *Publications of the ACOA,* 271; Whidden, "Importance of Recreation for Employees," *Publications of the ACOA,* 28–29, ACOA/AMI Records.

32. Fitch, "Birmingham District," 1540; "1912 Secretary-Treasurer's Annual Report," July 12, 1913, ACOA/AMI Records.

33. Figures on 1911 mine fatalities from Ward and Rogers, *Convicts, Coal, and the Banner Mine Tragedy,* 65. For a general report on the Palos explosion, see the *Birmingham Labor Advocate,* May 15, 1910.

34. *Birmingham Labor Advocate,* June 24, 1910, July 1, 1910.

35. Morrow developed a close and sometimes lucrative relationship with the ACOA. In 1919 the operators paid Morrow five thousand dollars for "services rendered in connection with legislative matters during the summer . . . at Montgomery." Morrow's "services" were more than likely connected to the operators' campaign to defeat a workmen's compensation bill pending before the state legislature. See "Executive Committee Meeting Minutes," Sept. 26, 1919, ACOA/AMI Records.

36. Ward and Rogers, *Convicts, Coal, and the Banner Mine Tragedy,* 67; *Birmingham Labor Advocate,* Feb. 3, 1911; Graebner, "Coal-Mine Operator and Safety," 498, 500; Ward and Rogers, *Convicts, Coal, and the Banner Mine Tragedy,* 69–70. On the ACOA's involvement in framing the bill and their acknowledgment that it met with "considerable opposition on the part of the United Mine Workers Organization [*sic*] of the district as well as others, and was only passed after continuous and hard work on the part of [the ACOA] Legislative Committee," see "1910 Secretary-Treasurer's Annual Report," June 6, 1911, ACOA/AMI Records. For a detailed account of the bill's navigation through the legislature at Montgomery, see Ward and Rogers, *Convicts, Coal, and the Banner Mine Tragedy,* 65–76. William Graebner has argued that mine safety reform at the national level was "motivated not by selfless humanitarianism but by the desire to beat the unions at their own game" and that, as in Alabama, legislation promoted by mine owners elsewhere in the United States "emphasized miner ignorance, carelessness, and disobedience as causes of mine accidents." See Graebner, "Coal-Mine Operator and Safety," 483–505.

37. Ward and Rogers, *Convicts, Coal, and the Banner Mine Tragedy,* 71, 75–76 (figures on the deaths at Banner on 25).

38. Fitch, "Birmingham District," 1531; *Birmingham Labor Advocate,* Nov. 24, 1911, May 13, 1910; Fitch, "Birmingham District," 1531. On Alabama mine fatality rates, see "Coal Mine Fatalities, 1911–1932," 1, in Department of Mines, State of Alabama, Erskine Ramsay Papers, Department of Archives and Manuscripts, Birmingham Public Library, Birmingham, Ala. This document shows that, with the exception of 1915, Alabama's death rate per ton exceeded the national average for every year between 1911 and 1921.

39. J deB. Hooper to Governor Emmett O'Neal, Apr. 17, 1911, in O'Neal Administrative Files, ADAH.

40. "Interview: Luther V. Smith"; H. W. Hamilton to Governor Emmett O'Neal, Feb. 26, 1912, O'Neal Administrative Files, ADAH.

41. *Birmingham Labor Advocate,* Jan. 10, 1913.

42. Ibid., Mar. 28, 1913.

43. On ACOA suspicions that the UMW was "holding the miners together secretly and . . . arranging to collect dues," see "Executive Board Minutes," Sept. 24, 1908, May 31, 1909. See also "1910 President's Annual Report" on outlay for "special services" (which included surveillance) in the amount of $6045.99. On their concern with the weights issue, see "Executive Committee Meeting Minutes," Mar. 28, 1913. All in ACOA/AMI Records.

44. Quote taken from the title of Ward and Rogers's chapter on mine safety legislation (*Convicts, Coal, and the Banner Mine Tragedy,* 65–76).

45. Hugh G. Grant to Milton Fies, Feb. 10, 1922, Milton Fies to Henry T. DeBardeleben, May 11, 1918, DeBardeleben Records.

46. See "The Coal Miners' Union during the War and Post-War" for an account of the

Montevallo strike (Taft Research Notes). The prominent black UMW organizer Walter Jones was raised at Aldrich. Along with his brother and his father, Jones played an active role in building the UMW local at Montevallo before being appointed a District 20 official.

47. "Message from Montevallo Mining Company President," Sept. 3, 1913, A. J. Jones et al. to Hon. Emmit O'Neal [*sic*], Jan. 6, 1914, O'Neal Administrative Files, ADAH. That the confrontation at Montevallo developed out of local grievances and was not forced by District 20 officials was acknowledged by W. S. Lovell, one of the new owners. Lovell complained to the district secretary J. L. Clemo that miners had "obstructed [new management] in every way," and "were only prevented from going on strike [in late November] by [union officals]." See Lovell to J. L. Clemo, Dec. 22, 1913, O'Neal Administrative Files, ADAH.

48. Henry T. DeBardeleben to Cyrus Garnsey Jr., June 5, 1915, Garnsey to DeBardeleben, June 12, 1915, G. F. Peter to DeBardeleben, June 7, 1915, DeBardeleben Records.

49. Fies to DeBardeleben, July 23, 1915, DeBardeleben Records. On McCormack's role in the fight at Montgomery, see *Montgomery Advertiser*, July 17, 1915, and *Birmingham Age-Herald*, July 17, 1915.

50. Rikard, "Experiment in Welfare Capitalism," 5.

51. Lewis, *Sloss Furnaces and the Rise of the Birmingham District*, 319. For an engaging discussion of the relation between architectural design and the cultivation of labor stability and loyalty, see Littmann, "Designing Obedience," 88–114. Birmingham district coal operators seem to have enjoyed one distinct advantage over the urban-based manufacturers discussed by Littman. The relative isolation of the coal camps and towns meant that, during the work week at least, miners could not travel back and forth between the workplace and towns outside of company control. Littman's argument that the rationale for comprehensive development of social amenities derived from employers' calculation that such facilities "might lure workers away from the coarse pleasures enjoyed in the surrounding working-class neighborhoods" nevertheless has important parallels in the mining district.

52. Tarbell, *Life of Elbert H. Gary*, 310–11.

53. On Edgewater, see Lewis, *Sloss Furnaces and the Rise of the Birmingham District*, 320. On labor stability, see U.S. Department of Labor, "Bulletin 263—Housing by Employers," 21, cited in Straw, "'This is Not a Strike, It Is Simply a Revolution,'" 204; Fitch, "Birmingham District," 1535.

54. Knowles, "Relation of the Operator to Sanitary and Social Improvements," 10; "Testimony of John Fitch," *U.S. Steel Hearings*, vol. 4, 2940.

55. Knowles, "Relation of the Operator to Sanitary and Social Improvements," 10–11; *Birmingham Age-Herald*, Jan. 28, 1907, Oct. 9, 1907. On the "pistol-toting evil" and its racial dimensions, see also the letter from the city coroner C. L. Spain in the *Birmingham Labor Advocate*, Oct. 10, 1913.

56. Bigelow, "Birmingham," 116.

57. Knowles, "Condensed Report and Recommendations on Sanitary Conditions," *Publications of the ACOA*, appendix B, ACOA/AMI Records; "Interview: Elmer Burton," 5; Knowles, "Condensed Report and Recommendations on Sanitary Conditions," introduction; *Birmingham Labor Advocate*, July 3, 1914.

58. "Interview: Annie Sokira," 13, "Interview: Ila Hendrix," 2, "Interview: Anne Sokira,"

10, Oral History Collection, Special Collections, Samford University, Birmingham, Ala. Montgomery argues that "the cohesiveness of the women in the mining towns" was one of three crucial factors that sustained the UMW, "in spite of numerous and often catastrophic defeats," as one of "the most influential organizations in the [early twentieth-century] labor movement" (*Fall of the House of Labor*, 333).

59. Montgomery, *Fall of the House of Labor*, 337. Their centrality in leading and maintaining above-ground community networks in Alabama catapulted mining camp women into prominent roles during industrial disputes, a point that is developed in some detail in Montgomery's discussion of the 1920 strike (271, 274–76).

60. Roden, "Value of Improved Sanitary and Living Conditions in Mining Camps," 34; Mrs. Gussie Moore in the *Birmingham Reporter*, June 19, 1918; Nolan, "Welfare Work of the Tennessee Coal, Iron, and Railroad Company," 274–5; Whidden, "Importance of Recreation for Employees," 29.

61. Lewis, *Sloss Furnaces and the Rise of the Birmingham District*, 320; Fies to Henry T. DeBardeleben, June 24, 1920, DeBardeleben Records.

62. Fitch, "Birmingham District," 1529–30; Cotter, *United States Steel*, 176; Bond, *Negro Education in Alabama*, 141–42; Fies, "Industrial Alabama and the Negro: Speech before the Alabama Mining Institute," Oct. 31, 1922, *Publications of the ACOA*, 49, ACOA/AMI Records. On paydays and the operation of camp commissaries, see Fitch, "Birmingham District," 1529–30; Harris, *Political Power in Birmingham*, 228; Fuller, "History of the Tennessee Coal, Iron, and Railroad Company," 111; Straw, "'This Is Not a Strike, It Is Simply a Revolution,'" 202–3, 209–10. On the coal companies and education in the Birmingham district generally, see Harris, *Political Power in Birmingham*, 168–74, Bond, *Negro Education in Alabama*, 226–61, and Dong, "Social and Working Lives of Alabama Coal Miners and Their Families," 56–58. Harris reports that TCI "found that one of its greatest problems was the lack of skilled labor in the South, where most of the population engaged in agriculture and lacked the skills, the discipline, and often the simple literacy necessary for productive industrial employment" (*Political Power in Birmingham*, 169).

63. Fitch, "Birmingham District," 1540; "Testimony of John Fitch," *U.S. Steel Hearings*, vol. 4, 2942; *Birmingham Labor Advocate*, June 30, 1917.

64. "DeBardeleben Oasis," 10; "Report of the President and Board of Directors of the AFICo for the Year 1916," AFICo Records; *Weekly Call*, Aug. 9, 1917; J. L. Westwood to Governor Kilby, Mar. 9, 1921, Governor Thomas Kilby Administrative Files, ADAH. The ACOA adopted similar procedures for processing workers' grievances. See "1914 President's Annual Report," 188, in ACOA/AMI Records.

65. UMW officials quoted in Taft, "The Coal Miners' Union During the War and Post-War," 3, 2, Taft Research Notes; *Birmingham Labor Advocate*, July 11, 1913 (see official UMW notice of same in this issue); Fies to DeBardeleben, ca. 1920, Fies to DeBardeleben, Feb. 23, 1914, DeBardeleben to Fies, Feb. 25, 1914, Fies to R. H. Franklin, June 13, 1920, DeBardeleben Records.

66. "Testimony of John Fitch," *U.S. Steel Hearings*, vol. 4, 2918–19.

67. *Birmingham Labor Advocate*, July 25, 1913. Interestingly, the UMW welfare proposal was developed by Ethel Armes, who in earlier years had been employed by the ACOA and had written a favorable portrait of the district's leading coal operators while under contract with the Chamber of Commerce.

68. Mitchell and Mitchell, *Industrial Revolution in the South,* 135; Spero and Harris, *Black Worker,* 247; *Birmingham Labor Advocate,* July 9, 1915.

69. Rikard, "George Gordon Crawford," 172.

Chapter 3: "Friends" in High Places

1. Bond, *Negro Education in Alabama,* 241. On American industry's increasing appreciation for the methods of welfare capitalism during this period, see Brandes, *American Welfare Capitalism,* 10–29.

2. Woodward, *Origins of the New South,* 208.

3. Green and Worthman argue that "while Southern industrialists and planters often grumbled about the 'lazy, shiftless, ignorant black laborer,' they contradicted themselves by expressing a clear preference for black workers. . . . By the early twentieth century, in fact, it was clear that blacks played an essential role in the south's rapid industrialization" ("Black Workers in the New South," 52). Davis et al., *Deep South,* 469; Wharton, *Negro in Mississippi,* 121.

4. Stampp, *Era of Reconstruction,* 80. On planter influence in late nineteenth-century mining, see Lewis, "Emergence of Birmingham as a Case Study of Continuity between the Antebellum Planter Class and Industrialization in the 'New South,'" 62–79, quote on 76.

5. *Birmingham Age-Herald,* Jan. 28, 1907; Fies, "Industrial Alabama and the Negro," 44, "Proceedings of the Institute Meeting of the Executive and Operating Officials of the Members of the Alabama Mining Institute," Oct. 31, 1922, ACOA/AMI Records.

6. *Birmingham Reporter,* July 29, 1916.

7. Washington, "Negro and the Labor Unions," 756; Jones, "The Negro in Labor and Industry: Address to the Tenth Annual NAACP Convention," in Fox and Meier (eds.), *Meetings of the Board of Directors, Records of Annual Conferences, Major Speeches, and Special Reports* (hereafter NAACP Papers, part 1).

8. *Birmingham Reporter,* July 28, 1917; "Interview: Earl Brown," B: 10–11, Working Lives Oral History Project, Hoole Library Special Collections, University of Alabama, Tuscaloosa. A black UMW member, Clarence Darden, recalled that union men usually boycotted such affairs: "Every once in a while . . . they'd have barbeque [*sic*] and things like that where the company could indoctrinate [the men], and . . . put out whatever propaganda they want to put out. . . . Of course my dad [who had been run out of Bibb County for union activity] and people like them, they would never go to them—they wouldn't go to the picnics and things." See "Interview: Clarence Darden," 9–10, Working Lives Oral History Project.

9. Sir George Campbell cited in Du Bois, *Black Reconstruction in America,* 590.

10. Norrell, *James Bowron,* 128.

11. "New Arrival," quoted in Dismuke and Norrell, *Other Side,* 2; "Interview: Rev. F. C. Jones with Cliff Kuhn," "Interview: John Garner with Cliff Kuhn," Working Lives Oral History Project; Lewis, *Black Coal Miners in America,* 67–68.

12. Franklin et al. to Bittner, Oct. 11, 1920, Frost to Bittner, Dec. 6, 1920, Bittner Papers.

13. Davis, *Labor and Steel,* 147; Worthman, "Black Workers and Labor Unions in Birmingham, Alabama," 376.

14. Ward and Rogers, *Labor Revolt in Alabama,* 41; Taft, *Organized Labor in American History,* 670; Foner, *Organized Labor and the Black Worker,* 79.

15. Henry T. DeBardeleben to Fies, June 18, 1914, DeBardeleben Records; *Birmingham Reporter*, Mar. 13, 1915; Fies to Henry T. DeBardeleben, Oct. 29, 1914, DeBardeleben Records.

16. Dr. F. L. Hoffman to Judge Gary, cited in Tarbell, *Life of Elbert H. Gary*, 311–12.

17. Lewis, *Black Coal Miners in America*, 27; Lichtenstein, *Twice the Work of Free Labor*, 75.

18. Lichtenstein, *Twice the Work of Free Labor*, 73.

19. On whipping licenses, see *Birmingham Labor Advocate*, June 6, 1913

20. Ward quoted in *Birmingham Labor Advocate*, Oct. 10, 1907; W. H. Hugh to Governor Thomas Kilby, Sept. 9, 1919, Kilby Administrative Files; Lewis, *Black Coal Miners in America*, 31; "A Dead Convict Don't Cost Nothin'," in Federal Writers' Project, *Alabama*, 407–9.

21. J. M. Austin to Hartwell Douglass (Alabama Convict Bureau), Aug. 21, 1913, "Convict Testimony—Huxford-Orvin Naval Stores," Oct. 11, 1913, O'Neal Administrative Files.

22. *Birmingham Labor Advocate*, May 10, 1907; Hart, *Social Problems of Alabama*, 44; "A Dead Convict Don't Cost Nothin'," in Federal Writers' Project, *Alabama*, 49; *Birmingham Labor Advocate*, June 11, 1915; J. M. Austin to Hartwell Douglass, Dec. 22, 1913, O'Neal Administrative Files.

23. J. M. Austin to Hartwell Douglass, Dec. 22, 1913, O'Neal Administrative Files; *Birmingham Labor Advocate*, Aug. 26, 1916; J. P. Hall to W. E. Matthews (State Convict Board), June 13, 1918, Governor Charles Henderson Administrative Files, ADAH.

24. Officials deplored "the tendency [in the camps] to the development of the grossest and most destructive forms of immorality" and did not distinguish between consensual and forced homosexual relations. Hart, *Social Problems of Alabama*, 51. See also "Interview: John Garner," B: 7–8, Working Lives Oral History Project.

25. Hart, *Social Problems of Alabama*, 46–47. County convicts are excluded from Hart's survey. *Birmingham Labor Advocate*, May 27, 1910, May 20, 1910; Hart, *Social Problems of Alabama*, 51.

26. U.S. Senate, Commission on Education and Labor, *Report of the Committee of the Senate upon the Relations between Labor and Capital*, 437–38; U.S. Immigration Commission, *Immigrants in Industries*, pt. 1, 218.

27. *Birmingham Labor Advocate*, Apr. 18, 1913.

28. *Birmingham Age-Herald*, Jan. 28, 1907; *Birmingham Labor Advocate*, Apr. 18, 1913.

29. *Birmingham Labor Advocate*, Apr. 18, 1913. On the operators' defense of the subcontracting system, see the testimonies of Ben Roden and the Sipsey superintendent Milton Fies before the Kilby Commission, Mar. 4, 1921, ACOA/AMI Records.

30. "Interview: Elmer Burton," 12; *Birmingham Labor Advocate*, Aug. 19, 1916, July 1, 1916; *United Mine Workers' Journal*, June 1, 1916.

31. *Birmingham Labor Advocate*, Aug. 5, 1916.

32. Fies to H. T. DeBardeleben, Oct. 1, 1914, DeBardeleben Records.

33. *Birmingham Age-Herald*, Jan. 28, 1907, Oct. 10, 1907; Anonymous, "Alabama Mining Camp," 790–91; Herndon, *Let Me Live*, 58.

34. Herndon, *Let Me Live*, 58; "Interview: Bobby Clayton," A: 2, Working Lives Oral History Project.

35. Meier, *Negro Thought in America*, 171; Gaines, *Uplifting the Race*, 20.

36. Smith had first won public attention during his efforts to organize a convention of

black Democrats at Montgomery in 1892. His role is recounted in Spero and Harris, *Black Worker,* 137; *Hot Shots,* Oct. 24, 1908, Aug. 26, 1908, July 30, 1908.

37. Washington, "Negro and his Relation to the Economic Progress of the South," 81–82. Stein argues that Washington "offered Southern leaders a way to combat black insurgency" and that his "promises of industrial peace were welcomed by capitalists throughout the country [who] sought Tuskegee graduates to man schools and manage black labor within industrial corporations. . . . They . . . needed Tuskegee's trained men and women to confront the militancy of an industrial proletariat" ("'Of Mr. Booker T. Washington and Others,'" 447–48). On Washington's relations with southern industrial employers, see also Woodward, *Origins of the New South,* 358–59; Cox, "Leadership among Negroes in the United States," 255–56; Meier, *Negro Thought in America,* 104–5.

38. Biographical information on Pettiford from Feldman, *Sense of Place,* 155. The relationship between TCI and the Washington/Pettiford venture in coal is discussed in a letter from Pettiford to Washington, Nov. 6, 1900, in Harlan (ed.), *Booker T. Washington Papers.* Hornady, *Book of Birmingham,* 69; "Address of Miss Kathryn M. Johnson to the Sixth Annual Conference of the NAACP," May 1914, and "Branches Organized during the Year of 1918," NAACP Papers, part 1. On Washington's influence in Birmingham, see also Straw, "'This Is Not a Strike, It Is Simply a Revolution,'" 190, and Worthman, "Black Workers and Labor Unions in Birmingham, Alabama," 382.

Significantly, the first NAACP chapter in Birmingham was established by Dr. Charles McPherson, a graduate of Atlanta University and an outsider among the city's accommodationist middle class. The organization apparently developed out of the wartime Colored Citizen's League, which had been more heavily influenced by the city's traditional black leadership. Established race leaders like Adams at first opposed McPherson's efforts but later accepted him and even supported the local chapter, a change that can be partly explained by their developing hostility toward Washington's successor at Tuskegee, R. R. Moton. The chapter lasted only two years, however. "In its initial stage," Feldman writes, "some supporters wanted a democratic organization and they tried to delegate responsibility to [working-class members]. But businessmen and professionals came to dominate it [and] unlike the national NAACP, [it] failed to take action against the racism and discrimination that penetrated most aspects of Birmingham society [and] adopted a low profile for fear of alienating any sympathetic whites" (*Sense of Place,* 270).

39. Spero and Harris, *Black Worker,* 364–35; Fies to DeBardeleben, Dec. 1, 1916, DeBardeleben Records; *Birmingham Reporter,* Jan. 31, 1920; Norrell, *James Bowron,* 243.

40. *Hot Shots,* July 30, 1908; *Birmingham Reporter,* Dec. 20, 1919.

41. Although the dearth of evidence precludes an in-depth treatment of internal division among Birmingham's black elite, the historian Joel Williamson's remarks about the "feudalization of black life" under Jim Crow are worth bearing in mind. Following Williamson, Fon Louise Gordon suggests in her important study, *Caste and Class: The Black Experience in Arkansas, 1880–1920,* that race institutions "produced an emerging black nobility—leaders who offered protection and commanded loyalty in return. These leaders contended with anyone who challenged their hegemony, including other blacks who tried to siphon off already inadequate resources." "Instead of harnessing the community's resources for a demonstration of solidarity against the hostile environment of Jim Crow," she writes, "the black community, led by the middle class, dissipated its power with

infighting and backbiting." See Gordon, *Caste and Class*, 103–4; Williamson, *Crucible of Race*, 53–57.

42. It was white workers' policy of racial exclusion—often codified in trade union charters—that handed over to employers a large potential strikebreaking force among African Americans. Whatley argues that "increasingly race-conscious African-American workers living in communities that cushioned them from the stigma of scabbing needed very little encouragement, especially when the striking union had a history of racial discrimination" ("African-American Strikebreakers from the Civil War to the New Deal," 549). Ginger makes clear that despite a long record of white hostility, black workers often rejected the strikebreaking role designated for them by employers, an observation born out in late nineteenth-century and early twentieth-century Alabama ("Were Negroes Strikebreakers?" 73–74). In any case, the ambivalence displayed by black workers toward unions that excluded them should be differentiated from the strategic vision advanced by industrial accommodationists, in which strikebreaking was advocated as a positive prescription for racial advance, as the means by which blacks could, in Oscar Adams's view, "command the laboring field." In the pre–World War I South, strikebreaking by blacks was explicitly linked to the accommodationist strategy of an alliance with ruling-class whites and acceptance of racial subordination and, as the example of the UMW illustrates, did not depend on a particular union's racial policy.

43. Harris, *Political Power in Birmingham*, 27; Meier, "Negro Class Structure and Ideology in the Age of Booker T. Washington," 258; *Birmingham Age-Herald*, Oct. 12, 1907; Hornady, *Book of Birmingham*, 69. On Adams, see Sledge, "Black Conservatism in the Twentieth-Century South," 4. On Pettiford, see Dismuke and Norrell, *Other Side*, 12, and Lewis, *Sloss Furnaces and the Rise of the Birmingham District*, 318. On Driver, see *Birmingham Reporter*, July 10, 1920. On Walker, see Meier, *Negro Thought in America*, 298.

44. In October 1917, for example, the *Reporter* protested that P. M. Edwards, the "wealthiest colored man in Birmingham," had been refused elevator service at the Jefferson County Courthouse. See *Birmingham Reporter*, Oct. 20, 1917.

45. Painter, "'Social Equality,' Miscegenation, Labor, and Power," 60. In his discussion of relations between black and white miners in the West Virginia coalfields, Joe Trotter makes the important point that "racial separation was not a static phenomenon: it was not entirely externally imposed by white racism, and it was not uniformly negative in its results." Blacks frequently preferred segregated, all-black institutions in which they enjoyed relative autonomy and self-government to formally integrated bodies dominated by whites. Trotter, *Coal, Class, and Color*, 50.

46. *Free Speech*, 1903, cited in Bailey, "In Their Own Voices," 63; *Birmingham Reporter*, Apr. 24, 1915. Others acknowledged that patronage of black businesses might involve financial sacrifice. A resolution adopted at the Fourth Atlanta University Conference in 1898, for example, echoed Burlong's insistence that "'The mass of the Negroes must learn to patronize business enterprises conducted by their own race, even at some slight disadvantage'" (cited in Frazier, *Black Bourgeoisie*, 155). Meier argues that "the doctrine of Negro support for Negro businesses could be employed as a rationalization for the Negro working classes to support Negro entrepreneurs who were basically interested in themselves rather than in the welfare of the masses of the race" ("Negro Class Structure and Ideology in the Age of Booker T. Washington," 259).

47. *Birmingham Reporter,* Dec. 4, 1920; *Hot Shots,* July 23, 1908, Apr. 1, 1911.

48. Aside from anecdotal accounts in company records, it is difficult to gauge how significantly conditions that black and white miners lived or worked under differed. Census data for 1910 shows that black mining women worked outside the home far more often than white women and that the whites most likely to employ domestic help were company superintendents, local merchants, physicians, and school teachers. At Carbon Hill, Piper, and Townley not a single white miner is listed as employing domestics of either race. Blacks also appear far more likely to have taken on boarders. According to the 1910 census, at Carbon Hill, Moses Grant (forty-seven), along with his wife and daughter (listed as laundresses), kept nine boarders in a rented house, all but two of whom worked in the mines (sheet 5). Jacob Oliver (thirty-two), a local barber originally from Mississippi, kept a (rented) boarding house along with his wife Hattie. Of six boarders, five worked in the mines (sheet 13). Of fifty-eight black families at Piper, forty either kept boarders or had extended family (or both) living with them. All forty lived in rented housing. See U.S. Department of Commerce and Labor, Bureau of the Census, *Thirteenth Census of the United States.*

49. For evidence of the orders' appeal within mining communities, see the listings of black fraternal orders in the *Birmingham Reporter,* Jan. 16, 1915. Feldman argues that "leading lodge members were often high profile church officials whose employment reflected an elevated status in the black community. While the working class comprised the bulk of the membership in most orders, elite African Americans rose to the top of the secret society hierarchy" (*Sense of Place,* 147). The black UMW veteran Earl Jones, however, recalled that "the local orders were one of the ways that word about organizing drives [and] union meetings were passed." "Interview: Cliff Kuhn with Earl Jones," Working Lives Oral History Project.

50. Meier, *Negro Thought in America,* 15; Anonymous, "Alabama Mining Camp," 791; *Birmingham Reporter,* May 17, 1919.

51. *Birmingham Reporter,* Dec. 4, 1920; National Negro Business League, "The New Program," ca. 1916, General Correspondence, box 40.268b, Robert Russa Moton Papers, Hollis Burke Frissell Library, Tuskegee University; *Birmingham Reporter,* Aug. 9, 1919.

52. Although the sources are scarce, a similarly lax atmosphere apparently prevailed in quarters occupied by whites. See "Interview: Bennie and Mary Amerson," Federal Writers' Project, *Alabama,* 496.

53. Gaines, *Uplifting the Race,* 94, 41; Sledge, "Black Conservatism in the Twentieth-Century South," 14–15; Anonymous, "Alabama Mining Camp," 790; "Interview: Bobby Clayton"; Anonymous, "Alabama Mining Camp," 791; Feldman, *Sense of Place,* 174; *Birmingham Reporter,* Aug. 12, 1916. Local tensions between female advocates of uplift and black mining camp women reflected a more general class divide among black women during this period. "Most black women were poor," writes Deborah Gray White in her evaluation of the National Association of Colored Women. "They lived lives weighed down by agricultural and domestic labor. Pervasive discrimination left them with few choices. They did not have time to read, much less to cultivate an appreciation for Beethoven," as uplifters prescribed. Moreover, she writes, the reformers' designation of chastity as the "litmus test of middle-class respectability . . . established an orthodoxy bound to drive a wedge between themselves and the masses of black women" (White, *Too Heavy a Load,* 78, 70).

54. *Birmingham Reporter*, Aug. 12, 1916; *Birmingham Labor Advocate*, July 1, 1917.

55. Feldman, *Sense of Place*, 170–71; *Birmingham Reporter*, Oct. 30, 1920; Feldman, *Sense of Place*, 177. Interestingly, this confrontation occurred in the midst of the 1920 coal strike, which had polarized the black community between UMW supporters and those loyal to the coal operators. It is impossible to draw any definite conclusions about whether there was any direct connection between these two developments.

56. Gordon, *Caste and Class*, 78; *Birmingham Reporter*, Aug. 9, 1919.

57. "Interview: Earl Jones," A: 6–7, Working Lives Oral History Project; *Hot Shots*, Aug. 26, 1908; *Birmingham Labor Advocate*, Sept. 4, 1908.

58. See Leighton, *Five Cities*, 125, and "Executive Board Minutes," Jan. 31, 1917, ACOA/AMI Records.

59. *United Mine Workers' Journal*, June 1, 1916; Flynt, "Alabama White Protestantism and Labor," 206, 208; *Birmingham Labor Advocate*, Apr. 5, 1919.

60. *Workmen's Chronicle*, Mar. 9, 1918; *Birmingham Reporter*, Jan. 10, 1920.

61. *Hot Shots*, Apr. 18, 1908; *United Mine Workers' Journal*, June 1, 1916;

62. U.S. Immigration Commission, *Immigrants in Industries*, 182.

Chapter 4: White Supremacy and Working-Class Interracialism

1. Tourgee, *Invisible Empire*, 62. In reality, organizations such as the Ku Klux Klan were often directed by "[white] men of considerable influence." On the social composition of the Reconstruction-era Klan, see Trelease, *White Terror*, xix, 51, 96, Tourgee, *Invisible Empire*, 37, 38, 54, 62, 79–80, 88, and Fitzgerald, *Union League Movement in the Deep South*, 200–233.

2. Woodward, *Origins of the New South*, 209; Bruce, *Rise of the New South*, 164–65; Wright, *Old South, New South*, 68; Ayers, *Promise of the New South*, 431.

3. Woodward, *Origins of the New South*, 211; Fields, "Ideology and Race in American History," 156–59.

4. Foner, *Policies and Practices of the American Federation of Labor*, 195–218, 282–307.

5. Ibid., 256–81, 219–55. On craft unionism and racial exclusion, see also Marshall, *Negro Worker*, 17–18, and Taft, *Organized Labor in American History*, 670. On the Socialist Party's record on the race question, see Foner, *American Socialism and Black Americans*, 238–53. On local AFL bodies' experiments with independent politics, see Greene, *Pure and Simple Politics*, and Strouthous, *U.S. Labor and Political Action*.

6. Woodward, *Origins of the New South*, 229. I am indebted to Eric Arnesen for pointing out to me that Herbert R. Northrup advanced an almost identical formulation almost a decade before Woodward. In testimony before the President's Commission on Fair Employment Practices in 1943, Northrup argued that the railroad brotherhoods were faced with the alternatives of either "admit[ting] the Negro into the union" to "present a solid front against the employer" or "forc[ing] the railroads to eliminate the Negro from train and engine service." Northrup, "Negro in the Railway Unions," 160.

7. *Bricklayer and Mason*, Sept. 1901; Foner, *Organized Labor and the Black Worker*, 103; Samuel Gompers to W. H. Paul, May 31, 1911, Taft Research Notes.

8. M. P. Carrick to Gompers, Mar. 5, 1903, Taft Research Notes.

9. *The Carpenter,* Sept. 9, 1903.

10. Ibid.

11. Lewis, *Black Coal Miners in America,* 45; *Birmingham Age-Herald,* Apr. 23, 1902, Apr. 24, 1902.

12. *Birmingham Labor Advocate,* Mar. 20, 1914.

13. Ibid., Oct. 23, 1908, Sept. 4, 1908.

14. Bond, *Negro Education in Alabama,* 25; Degler, *Other South,* 170–71.

15. Hyman, *Anti-Redeemers,* 191, 181.

16. Ibid., 191, 187. It was Gutman who first suggested the continuity between radical Republicanism and the Greenback-Labor Party. "Some southern blacks (and whites)," he wrote, "reacted to the decline of Radical enthusiasm and power by shifting to the Greenback-Labor Movement." He provides fragmentary evidence that this transfer of black political loyalty occurred in a number of places outside Alabama, including the coal regions of Pennsylvania, in Mississippi, West Texas, and Louisiana. See Gutman, "Black Coal Miners and the Greenback-Labor Party in Redeemer Alabama," 507–8.

17. Letwin suggests that while "no hard numbers concerning the racial make-up of the membership of the Greenback-Labor party survive . . . impressionistic data suggest that blacks figured prominently, and perhaps even predominated, among the Greenbacks, particularly in the coalfields" ("Race, Class, and Industrialization," 128). Gutman concurs that "blacks seemed more attracted to the movement than whites in Alabama" ("Black Coal Miners and the Greenback-Labor Party in Redeemer Alabama," 507).

18. Hyman, *Anti-Redeemers,* 182; Letwin, "Race, Class, and Industrialization," 7; Flynt, *Poor but Proud,* 98; Gutman, "Black Coal Miners and the Greenback-Labor Party in Redeemer Alabama," 511, 512.

19. Gutman, "Black Coal Miners and the Greenback-Labor Party in Redeemer Alabama," 509–10; Letwin, "Race, Class, and Industrialization," 121–23.

20. For a detailed reconstruction of the Knights' activity in the Alabama coalfields, see Letwin, "Race, Class, and Industrialization," 154–220. Quote from *Birmingham Age-Herald,* Sept. 2, 1895.

21. This account of the Patton strike relies heavily on Letwin's excellent interpretation. See Letwin, "Race, Class, and Industrialization," 186–91.

22. "Anonymous Referee's Report on 'Policing the "Negro Eden": Racial Paternalism in the Alabama Coalfields, 1908–21,'" (copy in author's possession); Hill, "Problem of Race in American Labor History," 199; Hill, "Rejoinder to Symposium on 'Myth-Making as Labor History,'" 593.

23. Harris, *Political Power in Birmingham,* 108; Letwin, "Race, Class, and Industrialization," 207.

24. *Birmingham Labor Advocate,* Nov. 12, 1909.

25. "On no known occasion [between 1878 and 1908]," Letwin writes, "did white miners mobilize to exclude African Americans as such from the mines." This is especially significant because it was during this period, when black miners still comprised a minority of the mining workforce, that an exclusionary policy would have been more practicable. See Letwin, "Interracial Unionism, Gender, and 'Social Equality' in the Alabama Coalfields," 532.

26. *Birmingham Labor Advocate,* Oct. 21, 1910, Nov. 18, 1910. Statham's allusion to Com-

er's "negro colony" is apparently a reference to the governor's midsized plantation in Comer, Alabama. While Comer employed white labor almost exclusively in his Avondale Mills, the largest textile operation in the state, black laborers worked his downstate plantation, as some had done under slavery. Statham's suggestion that these laborers enjoyed better treatment at the hands of Comer than did Birmingham district miners is badly mistaken, however. In August 1917, for example, Comer scolded his black female caretaker for distributing leftover milk and scraps of food among the field hands, insisting that it was "for the dogs and not for her to give away." In January 1923 he attempted to renege on bills for emergency medical treatment for some of his hands. And in March 1924, after a number of laborers had wandered off because of "dissatisfaction about their wages," Comer upbraided his overseer, insisting that "any negro who violates his contract" should be forced "to pay up before he leaves or beat him to death. I mean that literally." Statham's inability to find common cause with black field hands toiling under such conditions demonstrates how deeply his outlook had been corrupted by white supremacy. See Comer to W. B. Mitchell, Aug. 11, 1917; Comer to W. B. Mitchell, Jan. 13, 1923; and Comer to S. J. Dismuke, Mar. 3, 1924, all in Braxton Bragg Comer Papers, series 1a: Plantation Records, Southern Historical Collection, Louis Round Wilson Library, University of North Carolina at Chapel Hill.

27. *Birmingham Labor Advocate,* June 3, 1916.

28. Ibid., June 3, 1916, Jan. 24, 1913.

29. Reinforcing local motivations for interracial cooperation were the pressures felt by the national UMW to organize the Alabama fields. "It was . . . realized," according to Spero and Harris, "that if the coal from the competitive districts having a common market was to bring the same price, control would have to be established not only over differential advantages between these districts, but also over those in the districts that had different markets, and the coal of the latter might enter the markets of the former. . . . The invasion of the Lake [Superior] market by the coal from the southern fields made their unionization imperative if control of the central competitive field was to be maintained." Spero and Harris, *Black Worker,* 352–53.

30. *Birmingham Labor Advocate,* Jan. 15, 1915.

31. Letwin, "Race, Class, and Industrialization," 85; "J. R. Kennamer Speech at Oakman," Mar. 28, 1920, ACOA/AMI Records.

32. Ayers, *Promise of the New South,* 132; *Birmingham Labor Advocate,* Oct. 21, 1914, Nov. 13, 1914, July 8, 1916.

33. U.S. Immigration Commission, *Immigrants in Industries,* 200; Letwin, "Race, Class, and Industrialization," 103; "Interview: Luther V. Smith," 2; "Interview: Mr. And Mrs. William H. Walker," 12, Oral History Collection, Special Collections, Samford University, Birmingham, Ala. Making use of the surveillance reports filed by undercover agents during the 1894 strike, Alex Lichtenstein suggests that this camaraderie extended well beyond the workplace. See Lichtenstein, "Racial Conflict and Racial Solidarity in the Alabama Coal Strike of 1894."

34. "Green Fields Far Away," in Federal Writers' Project, *Alabama,* 441–65. "Rossia Cooker to Van Bittner," ca. 1920, Bittner Papers.

35. *Birmingham Labor Advocate,* Jan. 24, 1913, Aug. 5, 1916.

36. Ibid., Aug. 5, 1916.

37. Ibid.

38. Ibid., June 12, 1914, Aug. 5, 1916. The date is significant because it coincides with the reorganization of the UMW in the Birmingham district.

39. Lewis, *Black Coal Miners in America,* 45; *Birmingham Labor Advocate,* Aug. 15, 1903, Apr. 14, 1906; Straw, "'This Is Not a Strike, It Is Simply a Revolution,'" 141.

40. "Interview: Constance Jones-Price," 1A: 4–6, 2A: 1, "Interview: Rev. F. C. Jones," A: 7, Working Lives Oral History Project; "Affidavit Regarding Joe Sorsby," Feb. 12, 1921, Kilby Administrative Files.

41. "Dan J. DeJarnette to Van Bittner," Nov. 26, 1920, Dec. 13, 1920, Bittner Papers.

42. *Birmingham Labor Advocate,* Jan. 24, 1913; *United Mine Workers' Journal,* June 1, 1916; *Birmingham Labor Advocate,* July 1, 1916, June 17, 1916.

43. *United Mine Workers' Journal,* June 1, 1916; *Birmingham Labor Advocate,* June 17, 1916, July 1, 1916.

44. Lewis, *Black Coal Miners in America,* 45; 45–46; Roden, "Testimony before the Kilby Commission," Mar. 4, 1921, ACOA/AMI Records.

45. Lewis, *Black Coal Miners in America,* 46.

46. Ibid., 46.

47. *Birmingham Reporter,* July 28, 1917. Though it seems entirely plausible that incidents like this occurred in district locals, its prominent showcasing in the pages of Oscar Adams's *Reporter,* embedded among articles hostile to the UMW and advertisements from the leading coal companies, makes it suspect. Its credibility is also undermined by the account of the conversation that allegedly transpired between the newly elected black president and white miners. "I am a miner just like you," the candidate was supposed to have informed the whites, "only I have got the interest of *my company* and my people at heart." Such a statement seems more representative of Adams's outlook than that of black miners, particularly one recently elected union president.

48. John Williams to Milton Fies, Feb. 1922, DeBardeleben Records.

Chapter 5: War, Migration, and the Revival of Coalfield Militancy

1. "1914 President's Annual Report," ACOA/AMI Records.

2. *Birmingham Labor Advocate,* Sept. 15, 1913; "Report of Committee on Conditions in the Alabama Coal Mines," cited in Taft, "The Coal Miners Union during the War and Post-War," 2–3, Taft Research Notes.

3. *Birmingham Labor Advocate,* Mar. 14, 1913.

4. Taft, "The Coal Miners Union during the War and Post-War," 4, Taft Research Notes; *Birmingham Labor Advocate,* June 9, 1917; McCartin, "Labor's 'Great War,'" 149–50.

5. Foner, *Organized Labor and the Black Worker,* 130.

6. *Survey,* May 4, 1918.

7. "Early Surveys: Migration Study," Letters from Negro Migrants, 1916–17, 19, series 6, box 86, Urban League Papers, Manuscripts Division, Library of Congress, Washington, D.C.; *Survey,* Aug. 11, 1917; "Letters from Negro Migrants," 6, 17, Urban League Papers.

8. "Letters from Negro Migrants," 4, Urban League Papers; W. L. McMillen to R. L. Thornton, Nov. 2, 1916, straight numerical file 182363, Records of the Department of Justice, RG 60.

9. Tower, "Cotton Change in Alabama," 7; *Survey,* Feb. 27, 1915.

10. Tower, "Cotton Change in Alabama," 12.

11. *Montgomery Advertiser,* Aug. 24, 1916; "Early Surveys: Migration Study," Birmingham Summary, 2, "Migration Study: Newspaper Extracts, 1916–17," 2, Urban League Papers; *Birmingham Reporter,* Sept. 30, 1916.

12. Detailed accounts of this activity appear in the pages of the *Birmingham Reporter* throughout 1916 and 1917. Urban League investigators found that the American Cast Iron and Pipe Company's welfare work was directed mainly at checking emigration during this period. With thirteen hundred black and three hundred white employees, company officials reported in 1917 that they had "lost very few men through migration last year—the labor turnover was 120. This year 85." See "Early Surveys: Migration Study," Birmingham Summary, 4, Urban League Papers.

13. "Early Surveys: Migration Study," Birmingham Summary, 9, Urban League Papers; Fies to F. L. Davidson, Dec. 21, 1916, DeBardeleben Records; Fies cited in Letwin, *Challenge of Interracial Unionism,* 174.

14. Louis Harlan writes that Chisum was for a time Booker T. Washington's "most active spy" within William Monroe Trotter's organization in Boston and that "though Washington dropped him, Chisum continued spying for the rest of his life. During World War I he was a labor agent for northern manufacturers seeking cheap black labor and strikebreakers from southern farms. In the 1920s he served Robert R. Moton . . . as an undercover man [and in] the 1930s he worked secretly for the railroads and the Pullman Company against the organizing efforts of the Brotherhood of Sleeping Car Porters" (*Booker T. Washington,* 90, 93–94). Harlan was apparently unaware of the period during which Chisum worked for TCI—an appointment apparently secured for him by Moton through his long relationship with the company's management.

15. "Early Surveys: Migration Study," Birmingham Summary, 1, Urban League Papers; Melvin J. Chisum to R. R. Moton, July 12, 1918, General Correspondence, box 27.191, Moton Papers; *Birmingham Labor Advocate,* May 19, 1917.

16. Alex D. Pitts to Asst. Attorney General Samuel J. Graham, Oct. 30, 1916, straight numerical file 182363, Records of the Department of Justice, RG 60; "Early Surveys: Migration Study," Birmingham Summary, 1, Urban League Papers.

17. Pitts to Graham, Oct. 25, 1916, Pitts to Graham, Oct. 27, 1916, Robert N. Bell to Attorney General, Oct. 25, 1916, straight numerical file 182363, Records of the Department of Justice, RG 60.

18. Norrell, *James Bowron,* 244; *Birmingham Labor Advocate,* June 21, 1917, Oct. 7, 1916.

19. *Providence Journal,* Oct. 20, 1916; *Birmingham Reporter,* Sept. 30, 1916.

20. *Birmingham Reporter,* Aug. 5, 1916, Aug. 19, 1916; *Montgomery Advertiser,* Oct. 4, 1916.

21. "Migration Study: Newspaper Extracts, 1916–17," 26, Urban League Papers; *Birmingham Reporter,* Dec. 2, 1916, Oct. 6, 1917; *Montgomery Advertiser,* Dec. 11, 1916.

22. *Birmingham Reporter,* July 29, 1916, Sept. 30, 1916.

23. Ibid., Jan. 9, 1915, Jan. 22, 1916, Oct. 7, 1916; *Birmingham Age-Herald,* July 14, 1916, reprinted in *Birmingham Reporter,* July 22, 1916.

24. *Birmingham Reporter,* July 22, 1916, Nov. 18, 1916, Sept. 2, 1916.

25. Robert N. Bell to Attorney General, Oct. 25, 1916, straight numerical file 182363, Records of the Department of Justice, RG 60; *Birmingham Reporter,* Nov. 18, 1916.

26. "Early Surveys: Migration Study," Birmingham Summary, 5, 18, Urban League Papers.

27. *Birmingham Reporter,* Sept. 30, 1916; "Early Surveys: Migration Study," Birmingham Summary, 5, Urban League Papers; *Birmingham Reporter,* Nov. 18, 1916.

28. U.S. Attorney (Birmingham) to Attorney General, Oct. 28, 1916, straight numerical file 182363, Records of the Department of Justice, RG 60; "Early Surveys: Migration Study," Birmingham Summary, 5, Urban League Papers.

29. *Birmingham Reporter,* Dec. 23, 1916; Harry Pace, "The Business of Banking among Negroes," *Crisis* (Feb. 1927), 184–88, cited in Feldman, *Sense of Place,* 188; *Voice of the People,* Jan. 24, 1916, cited in Feldman, *Sense of Place,* 187; "Interview with Ellen Tarry," May 10, 1994, cited in Feldman, *Sense of Place,* 158; *Birmingham Reporter,* Jan. 22, 1916.

30. Taylor to Fies, Oct. 30, 1916, DeBardeleben Records; "Early Surveys: Migration Study," Birmingham Summary, 2, Urban League Papers.

31. "Early Surveys: Migration Study," Birmingham Summary, 2, Urban League Papers; *Survey,* Aug. 11, 1917; *Birmingham Reporter,* Sept. 30, 1916.

32. *Charleston News,* Dec. 17, 1916.

33. McCartin, "Labor's 'Great War,'" 146; "1914 President's Annual Report," 182, ACOA/AMI Records; McCartin, "Labor's 'Great War,'" 147; "Early Surveys: Migration Study," Birmingham Summary, 1, Urban League Papers.

34. Senator John H. Bankhead to George Gordon Crawford, Aug. 2, 1916, DeBardeleben Records. Bankhead's aim in the letter was to convince Crawford, as the head of TCI, to withdraw temporarily from the commercial coal market, calling it a "splendid time for your company to act as 'big brother' to the commercial coal companies."

35. *Birmingham Age-Herald,* Apr. 9, 1918.

36. Foner, *Labor and World War I,* 194.

37. George Haynes to Felix Frankfurter, Aug. 21, 1918, Records of the War Labor Policies Board, RG 1, National Archives, Washington, D.C.; "Annual Conference Proceedings, 1910–1950," NAACP Papers, part 1; Felix Frankfurter to George Haynes, Aug. 28, 1918, Records of the War Labor Policies Board, RG 1; Walter F. White to F. R. Belcher, Aug. 22, 1918, "Work or Fight Laws," in Fox and Meier (eds.), *Selected Branch Files* (hereafter NAACP Papers, part 12).

38. Walter White, "Notes on Work or Fight Laws," 1919, NAACP Papers, part 12.

39. "1918 Annual Report of the President," 15, ACOA/AMI Records; Bulletin no. 2, reprinted in the *Birmingham Reporter,* Dec. 15, 1917.

40. *Birmingham Reporter,* Nov. 17, 1917; Bulletin no. 10, "Annual Report for the Coal Mines," 1917, ACOA/AMI Records; *Birmingham Reporter,* Nov. 17, 1917.

41. Harris, *Political Power in Birmingham,* 201; *Birmingham Labor Advocate,* July 13, 1918. On the "Slacker's Cage," see also the *Birmingham Reporter,* July 13, 1918, which includes the text of an affidavit signed by the TCI timekeeper Albert Strong. Brown's offense seems to have been his stating that "War Savings Stamps and Thrift Stamps were a graft game and were no good, as they would not be redeemed by the Government." He reportedly complained that "there was constantly someone soliciting funds and that every day someone came trying to sell him a Liberty Bond, or asking something for the Red Cross and that he was tired of it, and would not give them anything."

42. On the 1918 steelworkers' strike, see McCartin, "Labor's 'Great War,'" 158–69, and

Northrup, "Negro and Unionism in the Birmingham, Alabama, Iron and Steel Industry," 27–40.

43. Quotes from Melvin Chisum to Secretary Halsey, Sept. 9, 1919, General Correspondence, box 40.268b, Moton Papers; *Workmen's Chronicle,* Mar. 16, 1918; "The Negro in Labor and Industry," NAACP Papers, part 1.

44. McCartin asserts that Birmingham provided the "first crucial testing ground of the NWLB. In no region of the nation had organized labor been less successful. In no region did the war offer workers a greater opportunity to redress past defeats and alter conditions of poverty and exploitation. In no region did the federal war labor agencies provide greater potential for disturbing business as usual." See McCartin, "Labor's 'Great War,'" 138.

45. Ibid., 163.

46. *Birmingham Labor Advocate,* Nov. 9, 1918; McCartin, "Labor's 'Great War,'" 179.

47. McCartin, "Labor's 'Great War,'" 168–69.

48. Ellis, "Federal Surveillance of Black Americans during the First World War," 1; Scheiber and Scheiber, "Wilson Administration and the Wartime Mobilization of Black Americans," 437; "Report of the Secretary," Dec. 1919, 1919 Annual Conference, NAACP Papers, part 1; "'Glasser File'—Memorandum for the Director of Military Intelligence," Aug. 15, 1919, Records of the Department of Justice, RG 60.

49. Murray, "Investigation: Alleged Fomenting Race Trouble," Oct. 13, 1919, Special Agent W. W., "Race Trouble—Birmingham," Feb. 17, 1920, W. W., "Race Trouble—Birmingham," ca. Jan. 1920, W. L. Hawkins, "Investigation: Alleged Fomenting of Race Trouble in Birmingham," Oct. 18–21, 1919, Birmingham File, Records of the Federal Bureau of Investigation, RG 65, National Archives, Washington, D.C.

50. Eaves, "Pro-Allied Sentiment in Alabama," 52–53; *Birmingham News,* Apr. 7, 1917; *New York Times,* Apr. 9, 1917, Apr. 7, 1917. On the accommodationists' cooperation in identifying black agitators to federal authorities, see the comments by Hollis B. Frissell, the principal of the Hampton School, in the *New York Times,* Apr. 13, 1917. See also the comments by George Haynes, the director of the federal government's Division of Negro Economics, cited in Scheiber and Scheiber, "Wilson Administration and the Wartime Mobilization of Black Americans," 449, and Special Agent W. W.'s comments on the Tuskegee Institute's "investigative work on radicalism" in "Race Trouble—Birmingham," Jan. 30, 1920, Birmingham File, Records of the Federal Bureau of Investigation, RG 65.

51. An inquiry by UMW officials revealed that Mathis had no formal connection to the federal government and that she was employed at the time by the Alabama Bankers' Association and apparently placed at the disposal of the ACOA. She was introduced to an integrated assembly of miners at Empire by the ACOA vice president Walter Moore, a prominent antiunion operator. See the *Birmingham Labor Advocate,* Nov. 24, 1917, which includes a letter from Nina C. Van Wrinkle of the United States Food Administration to District 20 President J. R. Kennamer, dated Nov. 18, 1917.

52. *Birmingham Age-Herald,* Sept. 3, 1917; Mrs. G. H. Mathis to Lewis [*sic*] Post, Sept. 3, 1917, "Controversy between the Alabama Coal Miners and Operators," Dispute Case Files, Records of the Federal Mediation and Conciliation Service, RG 280, National Archives, Washington, D.C.

53. Mrs. G. H. Mathis to Lewis [*sic*] Post, Sept. 3, 1917, Dispute Case Files, Records of

the Federal Mediation and Conciliation Service, RG 280; W. J. Boykin to Governor Charles Henderson, Sept. 28, 1917; K. O. Hudson to Governor Charles Henderson, July 30, 1917, Henderson Administrative Files, ADAH.

54. J. R. Kennamer to Department of Labor, Aug. 8, 1917; Dispute Case Files, "Alabama Coal Strike," Records of the Federal Mediation and Conciliation Service, RG 280; "Exhibit 'M': Evidence before the Kilby Commission," July 25, 1917, Kilby Administrative Files; "List of Men Permanently Dismissed," July 25, 1917, Bittner Papers; M. Gay to Department of Labor, Aug. 20, 1917, Chief Clerk's Files, "Special Problems—Birmingham," Records of the Department of Labor, RG 174, National Archives, Washington, D.C.

55. While workers welcomed the relief brought to the district by the U.S. Employment Bureau, Birmingham employers resisted efforts to extend its work into the postwar period. The *Labor Advocate* attributed their resistance to local employers' nostalgia for the private employment service that they had operated themselves prior to the war, under which they could "[refuse] employment to whom they choose, discriminating between union and non-union mechanics, effecting non-union agreements, allowing only such hours and wages as was necessary to keep their plants in operation, hiring scabs and strikebreakers during strikes and labor disputes. . . . These are the real reasons why the manufacturers of the Birmingham district . . . are asking for [the bureau's] abolishment." *Birmingham Labor Advocate*, Jan. 25, 1919.

56. McCartin writes that the "repercussions" from the NWLB's refusal to take action in the steelworkers' strike "were immediately felt in Alabama," "[breaking] the skilled workers' strike and halt[ing] the movement to organize the unskilled in its tracks. The loss of this central battle endangered the entire union movement in Alabama [and] substantially weakened, if it did not actually destroy the Board's authority throughout Alabama" ("Labor's 'Great War,'" 178).

57. The Justice Department's J. Reese Murray reportedly "hauled black union activists [in the steel dispute] into his offices to warn them that labor agitation was unpatriotic and to inform them that they would be closely watched" and "cooperated with . . . the Vigilantes [in] quashing the organizing drive among the steelworkers." He also directed the Justice Department's surveillance operations against Birmingham's black community. See ibid., 168, and Murray, "Investigation: Alleged Fomenting Race Trouble, Birmingham," Oct. 13, 1919, Records of the Federal Bureau of Investigation, RG 65.

58. The model developed by Labor Department officials for handling labor disputes in neighboring Mississippi illustrates the essential compatibility of interests between the federal government and the southern ruling class. In Bolivar County federal agents encouraged the founding of a Community Congress, made up of "five leading white planters and businessmen . . . and five leading Negro citizens of the County." The less formalized arrangement directed by industrial employers in the Birmingham district shared one important feature: the exclusion of black workers from any meaningful say in their treatment. See "Memorandum for the Director General of the Farm Service Reserve (Colored)," "Migration of Negroes North," Apr. 10, 1918, Chief Clerk's Files, Records of the Department of Labor, RG 174.

59. "1918 President's' Annual Report," 14, ACOA/AMI Records.

60. On employers' complaints about the Wilson administration's sympathetic relationship with organized labor, see the *New York Times*, Feb. 17, 1918. Quotes from *Birming-*

ham Labor Advocate, July 20, 1918; "President's Annual Report to the Stockholders of the Sloss-Sheffield Company," Mar. 1919, Sloss-Sheffield Steel and Iron Company Records, Department of Archives and Manuscripts, Birmingham Public Library, Birmingham, Ala.

61. "Interview: John Garner with Cliff Kuhn."

62. *Birmingham Labor Advocate,* Sept. 29, 1917, July 6, 1918. See also McCartin, "Labor's 'Great War,'" 150–53, and *Birmingham Labor Advocate,* July 28, 1917.

63. On prewar labor opposition and its collapse, see Foner, *Labor and World War I,* 40–77. For examples of antiwar opinion within Birmingham's labor movement, see *Birmingham Labor Advocate,* Oct. 16, 1914, Mar. 19, 1915. Quotes from Foner, *Labor and World War I,* 40; McCartin, "Labor's 'Great War,'" 145, 146; *Birmingham Labor Advocate,* Mar. 31, 1917.

64. *Birmingham Labor Advocate,* Mar. 31, 1917; *Weekly Call,* June 13, 1918.

65. Norrell, *James Bowron,* 233; Roden, "Testimony before the Kilby Commission," ACOA/AMI Records.

Chapter 6: "People Here Has Come to a Pass"

1. Hywel Davies, "Memorandum on the Salient Facts Relating to the Alabama Coal Strike," Records of the Federal Mediation and Conciliation Service, RG 280.

2. "The Facts about the Coal Strike in the Mineral District of Alabama and the Conditions Which Have Arisen from It," Kilby Administrative Files; "Statement of the President of the Sloss-Sheffield Steel and Iron Company to the Department of Labor," Alabama Coal Strike File, Dispute Case Files, Records of the Department of Labor, RG 174.

3. Erskine Ramsay to F. S. Peabody, June 14, 1917, Records of the Federal Mediation and Conciliation Service, RG 280.

4. *Birmingham Labor Advocate,* June 16, 1917; Mary E. McDowell to Henry Ward, Feb. 15, 1921, Bittner Papers; *Birmingham Labor Advocate,* June 23, 1917.

5. *Birmingham Labor Advocate,* June 16, 1917; *Birmingham Age-Herald,* Aug. 21, 1917; "To the Officers and Delegates to the Special Convention, District 20," July 31, 1917, Records of the Federal Mediation and Conciliation Service, RG 280; McCartin, "Labor's 'Great War,'" 150.

6. Taft Research Notes, 6–7; McCartin, "Labor's 'Great War,'" 153.

7. *Birmingham Labor Advocate,* Aug. 8, 1917; *Birmingham Reporter,* Aug. 18, 1917; *Manufacturers' Record,* cited in *Birmingham Reporter,* Aug. 18, 1917.

8. Taft Research Notes, 9.

9. *Birmingham Age-Herald,* Aug. 31, 1917; McCartin, "Labor's 'Great War,'" 154; Taft Research Notes, 8.

10. Taft Research Notes, 8.

11. Ibid.

12. Erskine Ramsay to F. S. Peabody, Aug. 14, 1917, Records of the Federal Mediation and Conciliation Service, RG 280. On the operators' refusal to sit down to negotiations with the UMW, see also "Statement of the President of Sloss-Sheffield to the Department of Labor," Aug. 16, 1917, Records of the Department of Labor, RG 174, and W. C. Liller to H. L. Kerwin, Jan. 10, 1921, Records of the Federal Mediation and Conciliation Service, RG 280.

13. Taft Research Notes, 9–10; McCartin, "Labor's 'Great War,'" 156; Roden, "Testimony before the Kilby Commission," 4, ACOA/AMI Records.

14. Southern Metal Trades Association, quoted in Newdick, "Employers Foster Race Prejudice," Feb. 24, 1919, Chief Clerk's Files, "Special Problems—Birmingham," Records of the Department of Labor, RG 174.

15. Taft Research Notes, 12.

16. *Birmingham Labor Advocate*, June 3, 1916; Newdick, "Employers Foster Race Prejudice," Feb. 24, 1919, Records of the Department of Labor, RG 174; J. C. Hartsfield to Governor Thomas E. Kilby, Sept. 13, 1919, Kilby Administrative Files.

17. Newdick, "Employers Foster Race Prejudice," Feb. 24, 1919, Records of the Department of Labor, RG 174.

18. H. P. Vaughn, "Memorandum for Mr. Haines [*sic*] on the Birmingham Race Situation," Mar. 5, 1919, Chief Clerk's Files, "Special Problems—Birmingham," Records of the Department of Labor, RG 174.

19. Vaughn, "Memorandum for Mr Haines [*sic*] on the Birmingham Race Situation," Records of the Department of Labor, RG 174.

20. *Birmingham Labor Advocate*, Mar. 6, 1920; Taft Research Notes, 14.

21. *Birmingham Labor Advocate*, July 17, 1920; "J. R. Kennamer Speech at Oakman," Mar. 28, 1920, ACOA/AMI Records.

22. J. W. Bridwell to H. L. Kerwin, July 20, 1920, Records of the Department of Labor, RG 174; *Birmingham Labor Advocate*, July 3, 1920; *Birmingham Age-Herald*, July 31, 1920.

23. *United Mine Workers' Journal*, June 1, 1920; *Birmingham News*, Sept. 5, 1920, cited in Straw, "United Mine Workers of America and the 1920 Coal Strike in Alabama," 109; *Birmingham News*, Sept. 2, 1920.

24. Bittner to P. Murray, Sept. 21, 1920; Bittner to Green, Sept. 30, 1920, Bittner Papers; *Birmingham News*, Sept. 24, 1920; Bittner to Green, Sept. 30, 1920, Bittner Papers.

25. Lewis, *Black Coal Miners in America*, 61; *Birmingham Reporter*, Oct. 9, 1920, Oct. 23, 1920.

26. *Birmingham Reporter*, Aug. 18, 1917, July 21, 1917, Sept. 11, 1920, Oct. 2, 1920, Oct. 30, 1920; Rameau to Kilby, Sept. 6, 1920, Kilby Administrative Files.

27. ACOA to Kilby, Sept. 6, 1920, in "For Our Coal Mine Employees: An Unqualified Statement," ACOA/AMI Records; *Birmingham Reporter*, Sept. 18, 1920.

28. *Birmingham Reporter*, Feb. 21, 1920; *United Mine Workers' Journal*, Oct. 1, 1920; *Birmingham Reporter*, Oct. 4, 1920.

29. "Van Bittner Speech," ca. Oct. 20, 1920, ACOA/AMI Records. The "Appeal to the Colored Mine Workers of Alabama" was published under the ostensible sponsorship of the "Association for the Welfare of Negroes in Alabama," but the union's adversaries charged, probably correctly, that the organization was little more than a UMW front. Copy in Bittner Papers.

30. Ellison to Kilby, June 22, 1920, July 18, 1920, Kilby Administrative Files.

31. *Birmingham News*, Sept. 19, 1920, cited in Straw, "United Mine Workers of America and the 1920 Coal Strike in Alabama," 116. For the operators' account of Adler's death, see the telegram from the Operators' Committee to Governor Kilby, Sept. 16, 1920, Kilby Administrative Files. The UMW's version of events appears in the *Birmingham Labor Advocate*, Oct. 9, 1920. "Transcript of Huddleston Speech," Sept. 19, 1920, in ACOA/AMI Records.

32. *Birmingham News,* cited in J. E. Ozborn to Kilby, Sept. 28, 1920, Milner Land Company to Kilby, Oct. 13, 1920, Kilby Administrative Files.

33. Burr to Kilby, July 17, 1920, Kilby Administrative Files.

34. *Birmingham Labor Advocate,* July 31, 1920; Denton to Kilby, Nov. 6, 1920, Kilby Administrative Files.

35. "Affidavit: McFadden et al.," Nov. 11, 1920, Kilby Administrative Files; "Affidavit: H. C. Jackson," Oct. 16, 1920, Bittner Papers.

36. Bittner to Murray, Oct. 6, 1920, Bittner Papers.

37. *Birmingham Labor Advocate,* Oct. 23, 1920; "Affidavit: R. B. Gilbert and Joe Thompson," ca. Oct. 1920, Kilby Administrative Files; *Birmingham Labor Advocate,* Oct. 30, 1920.

38. W. L. Branfield to Bittner, Sept. 27, 1920, Bittner Papers; J. S. Bolin to George A. Beauchamp, Oct. 8, 1920, Kilby Administrative Files; J. C. Walls to Bittner, Oct. 11, 1920, Bittner Papers; Sam Elliot to Kilby, Oct. 5, 1920, Kilby Administrative Files.

39. Bittner to Green, Sept. 30, 1920, Bittner Papers.

40. "The Facts about the Coal Strike," ca. Dec. 1920, Kilby Administrative Files; John D. McNell to President Wilson, Aug. 4, 1917, Records of the Department of Labor, RG 174.

41. *Birmingham News,* Sept. 5, 1920; Straw, "United Mine Workers of America and the 1920 Coal Strike in Alabama," 110; B. A. McWilliams to Bittner, Nov. 18, 1920, DeJarnette to Bittner, Nov. 26, 1920, Bittner Papers.

42. *Birmingham Reporter,* July 31, 1920; "Brief before the United States Coal Commission," Aug. 31, 1923, 57, ACOA/AMI Records.

43. J. Frost to Bittner, Dec. 6, 1920, W. T. Edge to Bittner, Jan. 6, 1920, Bittner Papers; R. F. Hancock to Kilby, Oct. 23, 1920, Kilby Administrative Files.

44. Blocton UMW to Bittner, Nov. 1920, Bittner Papers; R. F. Hancock to Kilby, Oct. 23, 1920, Kilby Administrative Files; G. W. Woodard to Bittner, Dec. 7, 1920, William Bradley to Bittner, Dec. 1920, J. Frost to Bittner, Jan. 3, 1921, Bittner Papers.

45. Bittner, "Press Statement," ca. Dec. 1920, "Speech of Van Bittner," ca. Oct. 1920, ACOA/AMI Records; Willie E. Meacham to Bittner, Feb. 3, 1921, J. G. Brown to Bittner, Nov. 26, 1920, R. W. Grubbs to Bittner, Nov. 1920, Bittner Papers.

46. Franklin et al. to Bittner, Oct. 11, 1920, Bittner Papers.

47. W. E. Creel to Bittner, Dec. 4, 1920, W. E. Creel to Bittner, Jan. 18, 1921, Bittner to J. H. Watkins, Jan. 28, 1921, H. D. Hill to Bittner, Dec. 19, 1920, B. A. McWilliams to Bittner, Nov. 18, 1920, Bittner Papers.

48. *Advance,* Dec. 4, 1920; Roden, "Testimony before the Kilby Commission," 17, ACOA/AMI Records.

49. *Advance,* Dec. 4, 1920.

50. Roden, "Testimony before the Kilby Commission," 17, ACOA/AMI Records; W. Cunningham to Bittner, Feb. 9, 1921, Bittner Papers.

51. "Special Report of the Grand Jury," Circuit Court of 10th Judicial Circuit of Alabama, 5, Kilby Administrative Files.

52. "Interview: Ellis Self," 9, Oral History Collection, Special Collections, Samford University, Birmingham, Ala.; W. Cunningham and Will Jones to Bittner, Jan. 30, 1921, W. L. Brasfield to Turnblazer, Jan. 10, 1921, Bittner Papers.

53. Elliot, *Autobiography of William T. Minor,* 24; Roden, "Testimony before the Kilby Commission," Mar. 21, 1921, ACOA/AMI Records.

54. Canon to Bittner, Dec. 23, 1920, Bittner Papers; Petition to Kilby, ca. Dec. 1920, Kilby Administrative Files; J. H. Watkins to Bittner, Jan. 8, 1920, Bittner Papers.

55. The UMW offered a five-thousand-dollar reward for "the arrest and conviction of the person or persons responsible" for Junius's murder. See the *Birmingham Labor Advocate,* Nov. 20, 1920.

56. "Statement of Facts," ca. Dec. 1920, Kilby Administrative Files; J. M. Henderson to Bittner, Dec. 27, 1920, Blocton UMW Local 3986 to Bittner, Dec. 15, 1920, Bittner Papers.

57. "Gentlemen: Address to the Kilby Commission," ca. Feb. 1921, "Affidavit Submitted to the Kilby Commission," Feb. 12, 1921, Kilby Administrative Files.

58. J. Frost to Bittner, Oct. 19, 1920, Bittner Papers.

59. Garnsey listed 132 white and 83 "colored" children; Straven showed 53 whites and 30 blacks. In January the Marvel local submitted separate requests for white and "colored" schoolbooks. This method of listing may have had more to do with logistics than racial consciousness. Perhaps the physical separation of black and white housing made it more practical for an individual of each race to calculate supplies in respective sections of the camp. Ploughman Harris to Bittner, Dec. 17, 1920, J. H. Wood to Bittner, Dec. 16, 1920, J. H. Watkins to Bittner, Jan. 13, 1921, Bittner Papers.

60. "Colord [*sic*] Union Woman" to Bittner, Dec. 6, 1920, R. W. White to Bittner, Nov. 26, 1920, Dan DeJarnette to Bittner, Dec. 13, 1920, Bittner Papers.

61. J. H. Wood to Bittner, Nov. 10, 1920, Green to Bittner, Nov. 6, 1920, T. J. Manning to Bittner, Dec. 28, 1920, Mrs. Lennie Creel to Bittner, Nov. 1920, Henry Woodford to Bittner, Dec. 10, 1920, W. E. Creel to Bittner, Jan. 18, 1921, B. W. Robinson to Bittner, Feb. 5, 1921, Bittner Papers.

62. *Birmingham Age-Herald,* Aug. 27, 1908; *Birmingham Labor Advocate,* Dec. 18, 1920.

63. "Speech of Van Bittner," ca. Oct. 1920, "Speech of Van Bittner," Nov. 7, 1920, ACOA/AMI Records. On the significance of the "social equality" charge and the explosive potency of gender, race, and sex in the Jim Crow South, see Painter, " 'Social Equality,' Miscegenation, Labor, and Power." On the salience of the "social equality" charge in the Alabama coalfields and its impact on the UMW's approach to race, see Letwin, "Interracial Unionism, Gender, and 'Social Equality' in the Alabama Coalfields."

64. Though it was apparently not known at the time either by miners or McDowell, Kilby was engaged in negotiations to purchase a mine during the course of the strike; whether he was pursuing this on his own behalf or on behalf of the state is unclear. See Robert Jemison to Kilby, Sept. 8, 1920, Kilby Administrative Files.

65. Bittner to Green, Oct. 4, 1920, McDowell to Ward, Feb. 15, 1921, Bittner Papers.

66. This commission included the former governor Charles Henderson, Judge James J. Mayfield, and George Denny, later to become president of the University of Alabama. Their work had a negligible effect on the strike, and Kilby appointed a second commission in the spring of 1921.

67. Straw, "United Mine Workers and the 1920 Alabama Coal Strike," 112–13; Liller to Kerwin, Nov. 29, 1920, Liller to Kilby, Jan. 29, 1921, Liller to Kerwin, Jan. 10, 1921, Records of the Federal Mediation and Conciliation Servive, RG 280. On Liller's forced withdrawal from the Alabama dispute, see letters from the Alabama Manufacturers' Association to Senator Oscar W. Underwood, Feb. 11, 1921, John H. Frye (Traders National Bank, Birmingham) to Underwood, Feb. 12, 1921, J. D. Kirkpatrick et al. to Underwood, Feb. 13, 1921,

Underwood to Secretary of Labor William B. Wilson, Feb. 17, 1921, all in "Birmingham Coal Strike," Dispute Case Files, Records of the Federal Mediation and Conciliation Service, RG 280.

68. W. T. Edge to Bittner, Jan. 6, 1921, "Our Harts Desire," ca. 1920–21, Bittner Papers.

69. Anonymous (Blocton) to Bittner, Nov. 1920, Bittner Papers, *Jasper Mountain Eagle*, ca. Jan. 30, 1921.

70. Jobie Jones to Bittner, Dec. 22, 1920, Alice Moore to Bittner, Nov. 24, 1920, Bittner Papers.

71. Oscar E. Fay to Bittner, Dec. 10, 1920, J. S. Long to Bittner, Feb. 2, 1920, Bittner Papers; *Birmingham Labor Advocate*, Feb. 19, 1921.

72. Straw, "United Mine Workers and the 1920 Alabama Coal Strike," 119; Liller to Kerwin, Jan. 10, 1921, Records of the Federal Mediation and Conciliation Service, RG 280; Liller to Bittner, Jan. 8, 1921, Bittner Papers.

73. Steiner to Kilby, Jan. 2, 1921, "Testimony Given by Conrad W. Austin at Jasper, Alabama," Jan. 25, 1921, Kilby Administrative Files; Liller to Kerwin, Jan. 10, 1921, Records of the Federal Mediation and Conciliation Service, RG 280.

74. Anonymous to Kilby, Jan. 9, 1921, "Petition to Kilby," Jan. 5, 1921, Kilby Administrative Files; J. H. Watkins to Bittner, Jan. 10, 1921, Bittner Papers. For a report on Klan activity during the strike, see "Trying to Run the Union Out of Alabama." Delegates from twenty-seven UMW locals meeting at Brookside in early April passed a resolution condemning the activity of the Klan, complaining that Klansmen had been "traversing [*sic*] the public high-ways in white robes and masked with license [plates] concealed." "Petition to Kilby," Apr. 8, 1921, Kilby Administrative Files.

75. A full account of the Baird incident appears in the *Advance*, Feb. 22, 1921.

76. UMW Statement, ca. Feb. 25, 1921, in Kilby Administrative Files; *Advance*, Feb. 22, 1921; *Jasper Mountain Eagle*, Feb. 8, 1921; *Advance*, Feb. 21, 1921.

77. Liller to Kerwin, Jan. 29, 1921, Records of the Federal Mediation and Conciliation Service, RG 280; Anonymous to Bittner, Feb. 12, 1921, Bittner Papers.

78. Bittner to Murray, Nov. 21, 1920, Bittner Papers; Straw, "United Mine Workers and the 1920 Alabama Coal Strike," 123.

79. "Decision of Governor Thomas Kilby," Mar. 19, 1921, Kilby Administrative Files.

80. Ibid.; *Birmingham Labor Advocate*, Mar. 25, 1921; "Annual Report of the Vice President and General Manager of AFICo for the Year 1921," Stockholders' Meetings, vol. 3, AFICo Records. Of the expenses incurred by AFICo during the strike, $56,000 went toward assessments to the ACOA while a further $6757.70 was reported as having been "expended for our own protection in the way of guns, extra deputies, etc."

81. Epigraph cited in a letter from Adamsville UMW to Governor Kilby, Apr. 4, 1921, I. G. Harden et al. to Kilby, Apr. 25, 1921, Kilby Administrative Files. J. F. Boston to Kilby, Dec. 28, 1920, Kilby Administrative Files; *Birmingham Labor Advocate*, Jan. 22, 1921; T. W. Halley to Bittner, Nov. 13, 1920, Bittner Papers.

82. "Speech of Van Bittner at Blocton," Mar. 30, 1921, Kilby Administrative Files.

83. Mrs. T. E. Reeves to Kilby, Apr. 2, 1921, Kilby Administrative Files.

84. Duncan et al. to Kilby, Apr. 8, 1921, Kilby Administrative Files; *Birmingham News*, Apr. 20, 1921; Mrs. Nona Bates to Kilby, Apr. 1, 1921, Kilby Administrative Files.

85. Committee Representing Miners of Winona Coal Company to Kilby, June 4, 1921,

Kilby Administrative Files; "Annual Report of the Vice President and General Manager of AFICo for the Year 1921," Stockholders' Meetings, vol. 3, "Executive Committee Minutes," Mar. 14, 1922, ACOA Subject Files: "Strike 1920–21," ACOA/AMI Records; Brownell, "Birmingham, Alabama," 48.

Conclusion

1. Baker, *Following the Color Line*, 18, 17.

2. "Resolution on Industrial Discrimination," NAACP Papers, part 1. On Smith's electoral campaign and black disfranchisement, see Woodward, *Tom Watson*, 220–42.

3. Bartley, *New South*, 66. Bartley writes that "To Myrdal, racism was a moral problem . . . rather than the product of socioeconoic and ideological forces. Racism was personal rather than institutional; it was a matter of individual psychology. . . . [He] put the ameliorative influence of education ahead of the reorganization of society and the redistribution of wealth and power [and] argued that the racism—the immorality—of working class whites was the central impediment to racial progress and that, among blacks, the 'theory of labor solidarity' was essentially 'escapist in nature' rather than reformist. The evidence suggested to him that African-American civic progress was dependent on the leadership of black elites and the cooperation of affluent whites rather than on the organized militancy of the economically disadvantaged" (*New South*, 67).

4. Litwack, *Trouble in Mind*, xiv.

5. Mathews, "Georgia 'Race Strike' of 1909," 617, 626–27.

6. Painter, "'Social Equality,' Miscegenation, Labor, and Power," 47.

Bibliography

Archival Sources

Alabama Department of Archives and History, Montgomery, Ala.
    Governor Braxton Bragg Comer Administrative Files
    Governor Charles Henderson Administrative Files
    Governor Emmett O'Neal Administrative Files
    Governor Thomas Kilby Administrative Files
Birmingham Public Library, Department of Archives and Manuscripts, Birmingham, Ala.
    Alabama Coal Operators' Association/Alabama Mining Institute Records
    DeBardeleben Coal Company Records
    Erskine Ramsay Papers
    Philip Taft Research Notes on Alabama Labor History, 1902–77
    Photographs: General Collection
    Republic Iron and Steel Company, Southern District: Photo Album
    Sloss-Sheffield Steel and Iron Company Records
Library of Congress, Manuscripts Division, Washington, D.C.
    Urban League Papers
National Archives, Washington, D.C.
    Records of the Coal Commission (RG 68)
    Records of the Department of Justice (RG 60)
    Records of the Department of Labor (RG 174)
    Records of the Federal Bureau of Investigation (RG 65)
    Records of the Federal Mediation and Conciliation Service (RG 280)
    Records of the War Labor Policies Board (RG 1)
Samford University, Special Collections, Birmingham, Ala.
    Oral History Collection
Tuskegee University, Hollis Burke Frissell Library, Tuskegee, Ala.
    Robert Russa Moton Papers

University of Alabama, Hoole Library Special Collections, Tuscaloosa, Ala.
  Records of the Alabama Fuel and Iron Company
  Working Lives Oral History Project
University of North Carolina, Louis Round Wilson Library, Southern Historical Collection, Chapel Hill, N.C.
  Braxton Bragg Comer Papers
West Virginia University, West Virginia and Regional History Collection, Morgantown, W.Va.
  Van Amburg Bittner Papers

Published Sources

Anonymous. "The Alabama Mining Camp." *Independent* 63 (Oct. 3, 1907): 790–91.

Armes, Ethel. *The Story of Coal and Iron in Alabama.* Birmingham: Birmingham Chamber of Commerce, 1910.

Arnesen, Eric. "Following the Color Line of Labor: Black Workers and the Labor Movement before 1930." *Radical History Review* 55 (1993): 53–87.

———. "Up from Exclusion: Black and White Workers, Race, and the State of Labor History." *Reviews in American History* 26:2 (March 1998): 146–74.

———. *Waterfront Workers of New Orleans: Race, Class, and Politics, 1863–1923.* New York: Oxford University Press, 1990.

Ayers, Edward L. *The Promise of the New South: Life after Reconstruction.* New York: Oxford University Press, 1992.

Bailey, Nancy Perrin. "In Their Own Voices: A Study of Birmingham's Weekly Press in the Progressive Era, 1900–1917." Master's thesis, University of Alabama, 1983.

Baker, Ray Stannard. *Following the Color Line: An Account of Negro Citizenship in the American Democracy.* New York: Doubleday, Page, and Co., 1908.

Bartley, Numan V. *The New South, 1945–1980.* Baton Rouge: Louisiana State University Press, 1995.

Berthoff, Roland T. "Southern Attitudes toward Immigration, 1865–1914." *Journal of Southern History* 17 (1951): 328–60.

Bigelow, Martha C. Mitchell. "Birmingham: Biography of a City of the New South." Ph.D. diss., University of Chicago, 1946.

Bond, Horace Mann. *Negro Education in Alabama: A Study in Cotton and Steel.* Reprint. Tuscaloosa: University of Alabama Press, 1994.

Bowron, James. "Convicts in Coal Mines in Alabama: An Explanation of the Methods of Handling Convicts in Coal Mining and Safeguarding Their Health as Presented by Col. Bowron." *Manufacturer's Record,* Aug. 9, 1923, 77–78.

Brandes, Stuart D. *American Welfare Capitalism, 1880–1940.* Chicago: University of Chicago Press, 1976.

Brier, Stephen. "In Defense of Gutman: The Union's Case." *Journal of Politics, Culture, and Society* 2:3 (Spring 1989): 382–95.

Brody, David M. "The Old Labor History and the New: In Search of an American Working Class." *Labor History* 20 (1979): 11–26.

Brownell, Blaine A. "Birmingham, Alabama: New South City in the 1920s." *Journal of Southern History* 38:1 (Feb. 1972): 21–48.

Bruce, Philip A. *The Rise of the New South.* Philadelphia: G. Barrie and Sons, 1908.

Callinicos, Alex. "Marxism and the Crisis in Social History" In *Essays in Historical Materialism.* Ed. John Rees. 25–40. London: Bookmarks, 1998.

Carlton, David L. "Paternalism and Southern Textile Labor: An Historiographical Review." In *Race, Class, and Community in Southern Labor History.* Ed. Gary M. Fink and Merl E. Reed. 17–26. Tuscaloosa: University of Alabama Press, 1994.

Carmer, Carl. *Stars Fell on Alabama.* New York: Farrar and Rinehart, 1934.

Cobb, James C. *Industrialization and Southern Society, 1877–1984.* Lexington: University Press of Kentucky, 1984.

———. *The Most Southern Place on Earth: The Mississippi Delta and the Roots of Southern Identity.* New York: Oxford University Press, 1992.

Corbin, David Alan. *Life, Work, and Rebellion in the Coal Fields: The Southern West Virginia Miners, 1880–1922.* Urbana: University of Illinois Press, 1981.

Cotter, Arundel. *United States Steel: A Corporation with a Soul.* Garden City, N.Y.: Doubleday, Page, and Co., 1921.

Cox, Oliver C. "Leadership among Negroes in the United States." In *Studies in Leadership: Leadership and Democratic Action.* Ed. Alvin W. Gouldner. 228–71. New York: Russell and Russell, 1950.

———. "The Leadership of Booker T. Washington." *Social Forces* 30 (1951): 91–97.

Cushing, George H. "The Birmingham Coal Puzzle." *American Coal Wholesaler* 104 (Oct. 4, 1919): 1–8.

Davis, Allison et al. *Deep South: A Social Anthropological Study of Caste and Class.* Reprint. Chicago: University of Chicago Press, 1988.

Davis, Horace B. *Labor and Steel.* New York: International Publishers, 1933.

"The DeBardeleben Oasis: Unionism's Last Frontier." *Alabama: Newsmagazine of the Deep South* 2 (1937): 9–10.

Degler, Carl. *The Other South: Southern Dissenters in the Nineteenth Century.* New York: Harper and Row, 1974.

Dismuke, Otis, and Robert J. Norrell. *The Other Side: The Story of Birmingham's Black Community.* Birmingham, Ala.: n.p., n.d. (copy at Schomburg Center for Research in Black Culture, Harlem, New York).

Dong, Jinsheng. "The Social and Working Lives of Alabama Coal Miners and Their Families, 1893–1925." Master's thesis, Auburn University, 1988.

Draper, Alan. *Conflict of Interests: Organized Labor and the Civil Rights Movement in the South, 1954–1968.* Ithaca, N.Y.: ILR Press, 1994.

Du Bois, W. E. B. *Black Reconstruction in America, 1860–1880.* Reprint. New York: Vintage Books, 1995.

Eaves, Robert Glenn. "Pro-Allied Sentiment in Alabama, 1914–1917: A Study of Representative Newspapers." *Alabama Review* 25:1 (Jan. 1972): 30–55.

Ellis, Mark. "Federal Surveillance of Black Americans during the First World War." *Immigrants and Minorities* 12:1 (Mar. 1993): 1–20.

Elliot, Carl Sr. *Autobiography of William T. Minor, Walker County, as told to Carl Elliot, Sr.* Jasper, Ala., 1977.

Elmore, Nancy. "The Birmingham Coal Strike of 1908." Master's thesis, University of Alabama, 1966.

Eskew, Glenn T. *But for Birmingham: The Local and National Movement in the Civil Rights Struggle.* Chapel Hill: University of North Carolina Press, 1997.

Federal Writers' Project. *Alabama: Life Histories.* M-3709. Southern Historical Collection, University of North Carolina at Chapel Hill. Microfilm.

Feldman, Glenn. "Research Needs and Opportunities: Race, Class, and New Directions in Southern Labor History." *Alabama Review* 51:2 (April 1998): 96–106.

Feldman, Lynne B. *A Sense of Place: Birmingham's Black Middle-Class Community, 1890–1930.* Tuscaloosa: University of Alabama Press, 1999.

Fickle, James E. "Management Looks at the 'Labor Problem': The Southern Pine Industry during World War I and the Postwar Era." *Journal of Southern History* 40:4 (Feb. 1974): 61–76.

Fields, Barbara J. "Race and Ideology in American History." In *Region, Race, and Reconstruction: Essays in Honor of C. Vann Woodward.* Ed. J. Morgan Kousser and James M. McPherson. 143–77. New York: Oxford University Press, 1982.

Fitch, John A. "Birmingham District: Labor Conservation." In "The Human Side of Large Outputs: Steel and Steel Workers in Six American States, Part IV." *Survey,* Jan. 6, 1912, 1527–40.

Fitzgerald, Michael W. *The Union League Movement in the Deep South: Political and Agricultural Change during Reconstruction.* Baton Rouge: Louisiana State University Press, 1989.

Flynt, Wayne. "Alabama White Protestantism and Labor, 1900–1914." *Alabama Review* 34:4 (Oct. 1981): 243–86.

———. *Poor but Proud: Alabama's Poor Whites.* Tuscaloosa: University of Alabama Press, 1989.

Foner, Philip S. *American Socialism and Black Americans: From the Age of Jackson to World War II.* Westport, Conn: Greenwood Press, 1977.

———. *Labor and World War I, 1914–1918.* Vol. 7 of *History of the Labor Movement in the United States.* New York: International Publishers, 1987.

———. *Organized Labor and the Black Worker, 1619–1973.* New York: International Publishers, 1974.

———. *The Policies and Practices of the American Federation of Labor, 1900–1909.* Vol. 3 of *History of the Labor Movement in the United States.* New York: International Publishers, 1964.

———. *Postwar Struggles, 1919–1920.* Vol. 8 of *History of the Labor Movement in the United States.* New York: International Publishers, 1988.

Fox, Mark, and August Meier, eds. *Meetings of the Board of Directors, Records of Annual Conferences, Major Speeches, and Special Reports, 1909–1950.* Part 1 of the NAACP Papers. Ann Arbor, Mich. 1981. Microfilm.

———, eds. *Selected Branch Files, 1913–1939,* series A: *The South.* Part 12 of the NAACP Papers. Ann Arbor, Mich. 1981. Microfilm.

Fox-Genovese, Elizabeth, and Eugene D. Genovese. *Fruits of Merchant Capital: Slavery and Bourgeois Property in the Rise and Expansion of Capitalism.* New York: Oxford University Press, 1983.

Frazier, E. Franklin. *Black Bourgeoisie: The Rise of a New Middle Class.* New York: Collier, 1962.

Fuller, Justin. "From Iron to Steel: Alabama's Industrial Evolution." *Alabama Review* 17:2 (April 1964): 137–48.

———. "History of the Tennessee Coal, Iron, and Railroad Company, 1852–1907." Ph.D. diss., University of North Carolina, 1966.

Gaines, Kevin K. *Uplifting the Race: Black Leadership, Politics, and Culture in the Twentieth Century.* Chapel Hill: University of North Carolina Press, 1996.

Garraty, John A. "The United States Steel Corporation versus Labor: The Early Years." *Labor History* 1:1 (1960): 3–38.

Gatewood, Willard B. "Aristocrats of Color, South and North: The Black Elite, 1880–1920." *Journal of Southern History* 54:1 (Feb. 1988): 3–20.

Genovese, Eugene. *Roll, Jordan, Roll: The World the Slaves Made.* New York: Vintage Books, 1972.

Gibbs, Henry Waight. "The Negro as an Economic Factor in Alabama." Ph.D. diss., Boston University, 1918.

Ginger, Ray. "Were Negroes Strikebreakers?" *Negro History Bulletin* 15 (Jan. 1952): 73–74.

Gordon, Fon Louise. *Caste and Class: The Black Experience in Arkansas, 1880–1920.* Athens: University of Georgia Press, 1995.

Grady, Henry W. "The South and Her Problems." In *The Complete Speeches and Orations of Henry W. Grady.* Ed. Edwin Dubois Shurter. 23–64. Austin, Tex.: Southwest Publishing Co., 1910.

Graebner, William. "The Coal-Mine Operator and Safety: A Study of Business Reform in the Progressive Period." *Labor History* 14:4 (Fall 1973): 483–505.

Green, James R., and Paul B. Worthman. "Black Workers in the New South, 1865–1915." In *Key Issues in the Afro-American Experience.* Ed. Nathan I. Huggins, Martin Kilson, and Daniel M. Fox. 47–69. New York: Harcourt Brace, 1971.

Greene, Julie. *Pure and Simple Politics: The AFL and Political Activism, 1881–1917.* Cambridge: Cambridge University Press, 1998.

Gutman, Herbert G. "Black Coal Miners and the Greenback-Labor Party in Redeemer Alabama: 1878–1879." *Labor History* 10:3 (Fall 1969): 506–35.

———. "The Negro and the United Mine Workers of America: The Career and Letters of Richard L. Davis and Something of Their Meaning, 1890–1900." In *The Negro and the American Labor Movement.* Ed. Jules Jacobsen. 49–127. New York: Anchor Books, 1968.

Halpern, Rick. *Down on the Killing Floor: Black and White Workers in Chicago's Packinghouses, 1904–54.* Urbana: University of Illinois Press, 1997.

———. "Organized Labor, Black Workers, and the Twentieth-Century South: The Emerging Revision." In *Race and Class in the American South since 1890.* Ed. Melvyn Stokes and Rick Halpern. 43–76. Oxford: Berg, 1994.

"Handling the Negro Miner in the South." *Coal Age,* May 30, 1914, 875.

Harlan, Louis R. *Booker T. Washington: The Wizard of Tuskegee, 1901–1915.* New York: Oxford University Press, 1983.

———, ed. *Booker T. Washington Papers.* Washington, D.C. 1981. Microfilm.

Harris, Carl V. *Political Power in Birmingham, 1871–1921.* Knoxville: University of Tennessee Press, 1977.

Harrison, Shelby M. "A Cash Nexus for Crime." *Survey,* Jan. 6, 1912, 1541–56.

Hart, Hastings H. *Social Problems of Alabama: A Study of the Social Institutions and Agencies of the State of Alabama as Related to Its War Activities.* Montgomery: n.p., 1918.

Herndon, Angelo. *Let Me Live.* New York: Random House, 1937.

Hill, Herbert. "Myth-Making as Labor History: Herbert Gutman and the United Mine Workers of America." *Journal of Politics, Culture, and Society* 2:2 (Winter 1988): 132–99.

———. "The Problem of Race in American Labor History." *Reviews in American History* 24:2 (1996): 189–208.

———. "Recent Effects of Racial Conflict on Southern Industrial Development." *Phylon* 20:4 (Winter 1959): 319–26.

———. "Rejoinder to Symposium on 'Myth-Making as Labor History: Herbert Gutman and the United Mine Workers of America.'" *Journal of Politics, Culture, and Society* 2:4 (Summer 1989): 587–95.

Honey, Michael K. *Black Workers Remember: An Oral History of Segregation, Unionism, and the Freedom Struggle.* Berkeley: University of California Press, 1999.

———. *Southern Labor and Black Civil Rights: Organizing Memphis Workers.* Urbana: University of Illinois Press, 1993.

Hornady, John R. *The Book of Birmingham.* New York: Dodd, Mead, and Co., 1921.

Horowitz, Roger. *"Negro and White, Unite and Fight!": A Social History of Meatpacking Unionism, 1930–90.* Urbana: University of Illinois Press, 1997.

Hyman, Michael R. *The Anti-Redeemers: Hill-County Political Dissenters in the Lower South from Redemption to Populism.* Baton Rouge: Louisiana State University, 1990.

Ignatiev, Noel. *How the Irish Became White.* New York: Routledge, 1995.

Jones, Jacqueline. *American Work: Four Centuries of Black and White Labor.* New York: W. W. Norton, 1998.

Kelly, Brian. "Policing the 'Negro Eden': Racial Paternalism in the Alabama Coalfields, 1908–21." *Alabama Review* 51:3 (July 1998): 163–83, and 51:4 (Oct. 1998): 243–65.

Leighton, George R. *Five Cities: The Story of Their Youth and Old Age.* New York: Harper and Brothers, 1939.

Letwin, Daniel L. *The Challenge of Interracial Unionism: Alabama Coal Miners, 1878–1921.* Chapel Hill: University of North Carolina Press, 1998.

———. "Interracial Unionism, Gender, and 'Social Equality' in the Alabama Coalfields, 1878–1908." *Journal of Southern History* 61:3 (Aug. 1995): 519–54.

———. "Race, Class, and Industrialization: Black and White Coal Miners in the Birmingham District of Alabama, 1878–1897." Ph.D. diss., Yale University, 1991.

Lewis, Ronald L. *Black Coal Miners in America: Race, Class, and Community Conflict, 1780–1980.* Lexington: University Press of Kentucky, 1987.

Lewis, W. David. "The Emergence of Birmingham as a Case Study of Continuity between the Antebellum Planter Class and Industrialization in the 'New South.'" *Agricultural History* 68:2 (Spring 1994): 62–80.

———. *Sloss Furnaces and the Rise of the Birmingham District: An Industrial Epic.* Tuscaloosa: University of Alabama Press, 1994.

Lichtenstein, Alex. "Racial Conflict and Racial Solidarity in the Alabama Coal Strike of 1894: New Evidence for the Gutman-Hill Debate." *Labor History* 36 (Winter 1995): 63–76.

———. *Twice the Work of Free Labor: The Political Economy of Convict Labor in the New South.* New York: Verso Press, 1996.

Limerick, Patricia Nelson. "Has 'Minority History' Transformed the Historical Discourse?" *Perspectives: Newsletter of the American Historical Association* 35:8 (Nov. 1997): 1, 32–36.

Littmann, William. "Designing Obedience: The Architecture and Landscape of Welfare Capitalism, 1880–1930." *International Labor and Working Class History* 53 (Spring 1998): 88–114.

Litwack, Leon F. *Trouble in Mind: Black Southerners in the Age of Jim Crow.* New York: Knopf, 1998.

Lynd, Staughton. "History, Race, and the Steel Industry." *Radical Historians' Newsletter* 76 (June 1997): 1–16.

Marshall, F. Ray. *The Negro Worker.* New York: Random House, 1967.

Marx, Karl. "The Eighteenth Brumaire of Louis Bonaparte." In Karl Marx and Frederick Engels, *Collected Works.* Vol 2. 99–197. New York: International Publishers, 1979.

Matthews, John Michael. "The Georgia 'Race Strike' of 1909." *Journal of Southern History* 40:4 (Nov. 1974): 613–30.

McCartin, Joseph. "Labor's 'Great War': American Workers, Unions, and the State, 1916–1920." Ph.D. diss., State University of New York at Binghamton, 1990.

McKiven, Henry M. Jr. *Iron and Steel: Class, Race, and Community in Birmingham, Alabama, 1875–1920.* Chapel Hill: University of North Carolina Press, 1995.

Meier, August. "Negro Class Structure and Ideology in the Age of Booker T. Washington." *Phylon* 23 (Fall 1962): 258–66.

———. *Negro Thought in America, 1880–1915: Racial Ideologies in the Age of Booker T. Washington.* Ann Arbor: University of Michigan Press, 1963.

Menzer, Mitch, and Mike Williams. "Images of Work: Birmingham, 1894–1937." *Journal of the Birmingham Historical Society* 7:1 (Jan. 1981): 10–13.

Merrill, Mike. "Interview with Herbert Gutman." In Herbert G. Gutman, *Power and Culture: Essays on the American Working Class.* Ed. Ira Berlin. 329–56. New York: Pantheon Books, 1987.

Minchin, Timothy J. *Hiring the Black Worker: The Racial Integration of the Southern Textile Industry, 1960–1980.* Chapel Hill: University of North Carolina Press, 1999.

Mitchell, Broadus, and George Sinclair Mitchell. *The Industrial Revolution in the South.* Baltimore: Johns Hopkins University Press, 1930.

Montgomery, David. *The Fall of the House of Labor: The Workplace, the State, and American Labor Activism, 1865–1925.* Cambridge: Cambridge University Press, 1987.

Myrdal, Gunnar. *An American Dilemma: The Negro Problem and Modern Democracy.* New York: Harper and Brothers, 1944.

Nelson, Bruce. "Class, Race, and Democracy in the CIO: The 'New' Labor History Meets the 'Wages of Whiteness.'" *International Review of Social History* 41 (1996): 351–74.

Norrell, Robert J. *James Bowron: The Autobiography of a New South Industrialist.* Chapel Hill: University of North Carolina Press, 1991.

Northrup, Herbert R. "The Negro and Unionism in the Birmingham, Alabama, Iron and Steel Industry." *Southern Economic Journal* 10 (July 1943): 27–40.

———. "The Negro in the Railway Unions." *Phylon* 5 (1944): 160.

Norwood, Stephen H. "Bogalusa Burning: The War against Biracial Unionism in the Deep South, 1919." *Journal of Southern History* 63:3 (Aug. 1997): 591–628.

Obadele-Starks, Ernest. "Black Labor, the Black Middle Class, and Organized Protest along the Upper Texas Gulf Coast, 1883–1945." *Southwestern Historical Quarterly* 103:1 (July 1999): 53–65.

Painter, Nell Irvin. "'Social Equality,' Miscegenation, Labor, and Power." In *The Evolution of Southern Culture.* Ed. Numan V. Bartley. 47–67. Athens: University of Georgia Press, 1988.

Preston, William Jr. *Aliens and Dissenters: Federal Suppression of Radicals, 1903–33.* Cambridge: Harvard University Press, 1963.

Reich, Stephen A. "Soldiers of Democracy: Black Texans and the Fight for Citizenship, 1917–1921." *Journal of American History* 82:4 (March 1996): 1478–1504.

Richards, Paul David. "Racism in the Southern Coal Industry, 1890–1910." Master's thesis, University of Wisconsin at Madison, 1969.

Rikard, Marlene Hunt. "An Experiment in Welfare Capitalism: The Health Care Services of the Tennessee Coal, Iron, and Railroad Company." Ph.D. diss., University of Alabama, 1983.

———. "George Gordon Crawford: Man of the New South." *Alabama Review* 31:3 (July 1978): 163–81.

Roediger, David R. "'Labor in White Skin': Race and Working-Class History." In *Towards the Abolition of Whiteness: Essays on Race, Politics, and Working Class History.* 21–38. New York: Verso, 1994.

———. *Towards the Abolition of Whiteness: Essays on Race, Politics, and Working Class History.* New York: Verso, 1994.

———. *The Wages of Whiteness: Race and the Making of the American Working Class.* New York: Verso, 1991.

———. "What If Labor Were Not White and Male? Recentering Working-Class History and Reconstructing Debate on the Unions and Race." *International Labor and Working Class History* 51:2 (Spring 1997): 72–95.

Saville, John. "The Radical Left Expects the Past to Do Its Duty." *Labor History* 18:2 (Spring 1977): 267–74.

Scheiber, Janet Lang, and Harry N. Scheiber. "The Wilson Administration and the Wartime Mobilization of Black Americans, 1917–18." *Labor History* 10:3 (Fall 1969): 433–58.

Sitton, Thad, and James H. Conrad. *Nameless Towns: Texas Sawmill Communities, 1880–1942.* Austin: University of Texas Press, 1998.

Sledge, James L. "Black Conservatism in the Twentieth-Century South." Paper delivered at the Southern Conference on Afro-American History, Feb. 1992.

Spero, Sterling D., and Abram L. Harris. *The Black Worker: The Negro and the Labor Movement.* New York: Columbia University Press, 1931.

Stampp, Kenneth. *The Era of Reconstruction: 1865–77.* New York: Knopf, 1965.

Stein, Judith. "'Of Mr. Booker T. Washington and Others': The Political Economy of Racism in the United States." *Science and Society* 38:4 (Winter 1974–75): 422–53.

———. *Running Steel, Running America: Race, Economic Policy, and the Decline of Liberalism.* Chapel Hill: University of North Carolina Press, 1998.

———. *The World of Marcus Garvey: Race and Class in Modern Society.* Baton Rouge: Louisiana State University Press, 1986.

Straw, Richard A. "'This Is Not a Strike, It Is Simply a Revolution': Birmingham Miners Struggle for Power, 1894–1908." Ph.D. diss., University of Missouri, 1980.

———. "The United Mine Workers of America and the 1920 Coal Strike in Alabama." *Alabama Review* 48:2 (April 1975): 104–28.

Strouthous, Andrew. *U.S. Labor and Political Action: A Comparison of Independent Political Action in New York, Chicago, and Seattle.* New York: Macmillan, 1999.

Taft, Philip. *Organized Labor in American History.* New York: Harper and Row, 1964.

———. *Organizing Dixie: Alabama Workers in the Industrial Era.* Rev. and ed. Gary M. Fink. Westport, Conn.: Greenwood Press, 1981.

Tarbell, Ida M. *The Life of Elbert H. Gary: The Story of Steel.* New York: D. Appleton, 1925.

Thompson, Edward P. *Customs in Common: Studies in Traditional and Popular Culture.* New York: New Press, 1993.

Tourgee, Albion W. *The Invisible Empire.* Reprint. Baton Rouge: Louisiana State University Press, 1989.

Tower, J. Allen. "Cotton Change in Alabama, 1879–1946." *Economic Geography* 26:1 (Jan. 1950): 6–28.

Trelease, Allen W. *White Terror: The Ku Klux Klan Conspiracy and Southern Reconstruction.* Baton Rouge: Louisiana State University Press, 1971.

Trotter, Joe William. "African-American Workers: New Directions in U.S. Labor Historiography." *Labor History* 36 (Summer 1995): 495–523.

———. *Coal, Class, and Color: Blacks in Southern West Virginia, 1915–1932.* Urbana: University of Illinois Press, 1990.

———, ed. *The Great Migration in Historical Perspective: New Dimensions of Race, Class, and Gender.* Bloomington: Indiana University Press, 1991.

"Trying to Run the Union Out of Alabama." *United Mine Workers' Journal,* Apr. 15, 1921, 13.

U.S. Department of Commerce and Labor, Bureau of the Census. *Thirteenth Census of the United States: Population.* 1910.

———. *Fourteenth Census of the United States: Population.* 1920.

U.S. Department of Labor, Bureau of Labor Statistics. *History of Wages in the United States from Colonial Times to 1928,* bulletin no. 604. 1934. Reprint 1966.

U.S. Immigration Commission. *Immigrants in Industries: Bituminous Coal Mining,* vol. 2. 61st Congress, 2d session. 1911.

U.S. Senate, Commission on Education and Labor. *Report of the Committee of the Senate upon the Relations between Labor and Capital,* vol. 4. 1885.

U.S. Senate, Committee on Investigation of the United States Steel Corporation. *Hearings on United States Steel Corporation.* 8 vols. 1912.

Walker, Clarence E. *Deromanticizing Black History: Critical Essays and Reappraisals.* Knoxville: University of Tennessee Press, 1991.

Ward, Robert David, and William Warren Rogers. *Convicts, Coal, and the Banner Mine Tragedy.* Tuscaloosa: University of Alabama Press, 1987.

———. *Labor Revolt in Alabama: The Great Strike of 1894.* Tuscaloosa: University of Alabama Press, 1965.

Washington, Booker T. "The Negro and His Relation to the Economic Progress of the

South." In *Selected Speeches of Booker T. Washington.* Ed. E. Davidson Washington. Garden City, N.Y.: Doubleday, Doran, and Co., 1932. 78–86.

———. "The Negro and the Labor Unions." *Atlantic Monthly* 111:6 (June 1913): 756.

Wharton, Vernon Lane. *The Negro in Mississippi, 1865–1890.* Chapel Hill: University of North Carolina Press, 1947.

Whatley, Warren C., "African-American Strikebreakers from the Civil War to the New Deal." *Social Science History* 17 (Winter 1993): 525–58.

White, Deborah Gray. *Too Heavy a Load: Black Women in Defense of Themselves, 1884–1994.* New York: Norton, 1999.

Wiebel, Arthur V. *Biography of a Business.* Birmingham: United States Steel (TCI Division), 1960.

Williamson, Joel. *The Crucible of Race: Black-White Relations in the American South since Emancipation.* New York: Oxford University Press, 1984.

Wilson, Bobby M. "Structural Imperatives behind Racial Change in Birmingham, Alabama." *Antipode* 24:3 (1992): 171–202.

Wingerd, Mary Lethert. "Rethinking Paternalism: Power and Parochialism in a Southern Mill Village." *Journal of American History* 83:3 (Dec. 1996): 872–902.

Woodward, C. Vann. *Origins of the New South, 1877–1913.* Baton Rouge: Louisiana State University Press, 1971.

———. *Tom Watson: Agrarian Rebel.* New York: Macmillan, 1938.

Worthman, Paul B. "Black Workers and Labor Unions in Birmingham, Alabama, 1897–1904." *Labor History* 10:3 (Summer 1969): 375–407.

———. "Working Class Mobility in Birmingham, Alabama, 1880–1914." In *Anonymous Americans: Explorations in Nineteenth-Century Social History.* Ed. Tamara K. Haveren. 172–213. Englewood Cliffs, N.J.: Prentice Hall, 1971.

Wright, Gavin. *Old South, New South: Revolutions in the Southern Economy since the Civil War.* New York: Basic Books, 1986.

Index

BRIAN KELLY is a lecturer in American history at the Queen's University of Belfast in Northern Ireland. He is working on a project on black elites and the labor question in the Jim Crow South.

The Working Class in American History

University of Illinois Press
1325 South Oak Street
Champaign, IL 61820-6903
www.press.uillinois.edu